Universitext

José María Montesinos

Classical Tessellations and Three-Manifolds

With 225 Figures Including 29 Colour Illustrations

Springer-Verlag
Berlin Heidelberg New York
London Paris Tokyo

José María Montesinos-Amilibia
Facultad de Matemáticas
Universidad Complutense
28040 Madrid, Spain

Mathematics Subject Classification (1985):
57N10, 51M20, 05B45, 57M05, 57M12, 57M25, 51F15, 51M05,
51M10, 20F38, 52A25, 52A45

ISBN-13:978-3-540-15291-0   e-ISBN-13:978-3-642-61572-6
DOI:10.1007/978-3-642-61572-6

Library of Congress Cataloging in Publication Data
Montesinos, José, 1944- Classical tessellations and three-manifolds. (Universitext)
Bibliography: p. 1. Tessellations (Mathematics) 2. Three-manifolds (Topology) I. Title.
QA166.8.M66   1987   514'.223   87-20645

2141/3140-543210

A mis padres Lorenzo y Victoria,
y en memoria de mi amigo Andrés.

Preface

"Más has dicho, Sancho, de lo que sabes (dixo Don
Quixote), que hay algunos que se cansan en saber,
y averiguar cosas que después de sabidas, y
averiguadas, no importa un ardite al entendimiento,
ni a la memoria."

*"You have said more than you know, Sancho", said
Don Quixote, "for there are some who tire themselves out learning and proving things which, once
learnt and proved, do not concern either the understanding or the memory a jot."*
Cervantes, Don Quixote, Part II, Chapter LXXV, Of
the great Adventure of Montesinos' Cave in the
heart of La Mancha, which the valorous Don Quixote
brought to a happy ending.

This book explores a relationship between classical tessellations
and three-manifolds.

All of us are very familiar with the symmetrical ornamental motifs
used in the decoration of walls and ceilings. Oriental palaces
contain an abundance of these, and many examples taken from them
will be found in the following pages. These are the so-called
mosaics or symmetrical tessellations of the *euclidean plane*. Even
though we can imagine or even create very many of them, in fact the
rules governing them are quite restrictive, if our purpose is to
understand the symmetric group of the tessellation, that is to say,
the group consisting of the plane isometries which leave the tessellation invariant. From this point of view, it can be proved that
there are precisely seventeen possible groups of symmetry for
euclidean-plane tessellations, discounting those tessellations which
can be considered as extensions of linear tessellations.

On the surface of a sphere, i.e. on the *spherical plane*, the
possible symmetrical motifs have the same groups of symmetry as the
platonic solids inscribed in the sphere. Assuming the bipyramid of

regular base with an arbitrary number of sides to be included among
the platonic solids, we remark that in the spherical case, infinitely
many groups of symmetry are possible. However, these groups can be
divided into a finite number of families. The most representative
examples of these groups of spherical symmetry are found in the
mineral kingdom. Internally a mineral can be thought of as an
aggregate of atoms arranged in an orderly pattern and constituting
a tessellation of three dimensional euclidean space. This structure
can be discerned by modern technical procedures and its group of
symmetry is called *internal*. What *we* see - macroscopically - is a
crystal. We can consider this crystal as a growth around a point,
in which case its group of symmetry is the subgroup consisting of the
elements of the internal group of symmetry which fix the point. More
precisely, this subgroup is the group of symmetry of the *crystal
germ*, or if one prefers, of a crystal in its first instant of
growth. However, it must be said that in most cases, the macrocrystal
will suffer an increase or a decrease (or indeed a total dis-
appearance) of some of its faces, due to the effect of the subgroup
of translations of the internal structure. If we keep this reser-
vation in mind, then mineralogy provides us with the most beautiful examples of spherical symmetry, since the "crystal germ" group,
being a subgroup of the group of isometries of euclidean space
fixing a point, is also a group of isometries of the sphere.

In the *hyperbolic plane* we also find tessellations but in much
greater variety than either the euclidean or the spherical plane.
It is sufficient to observe that given three angles whose sum is
less than π, a hyperbolic triangle exists having these angles. To
be sure the lesser the sum of the angles, the greater the area of
the triangle will be. If these angles are integral fractions of π,
the kaleidoscope whose mirrors form that triangle, will reflect a
tessellation of the hyperbolic plane. Moreover, the isotropy group
of a vertex will be the dihedral group of 2α elements, if α is
that fraction of π which defines the angle in that vertex. By vary-
ing α we obtain differing tessellations. It does not seem likely that
examples of such tessellations will be found in Nature. It is
again in the sphere of Art that the most interesting examples can be
found and thus, many of Escher's drawings are tessellations of the
hyperbolic plane.

In this book we deal with euclidean, spherical and hyperbolic
tessellations. More specifically, we study the *space of positions*
of these tessellations. Let us consider an example: take the
spherical pattern formed by a platonic solid, for instance, a
regular dodecahedron inscribed in a sphere. Keeping the sphere
fixed, we move the dodecahedron rigidly, forcing it to adopt all
possible inscribed positions in the sphere. We do not differentiate
between the parts of the dodecahedron and hence a particular position
is defined by the dodecahedron itself (occupying *that* particular
position of space and no other).

We have then a topological space whose points are the positions of
the dodecahedron, with some natural topology. It turns out that
this topological space is in fact an orientable, closed three-
manifold; and a very well-known one indeed, since it is homeomorphic
to the first homology sphere ever known. This homology sphere was
discovered by Poincaré in his celebrated 1904 paper entitled
"Cinquième complément à l'analysis situs" and published by Rendi-
conti del Circolo Matematico di Palermo (this same paper contains
the question known today as the Poincaré conjecture).

Poincaré was at the time investigating the depth of the homology
groups of a manifold. In an earlier work he had conjectured that a
three-manifold having the same homology groups as the three-sphere,
is the three-sphere. He disproved this by constructing the *Poincaré
homology sphere* mentioned above. He gave a Heegaard diagram of the
manifold, checked that it had trivial homology and proved that the
fundamental group of the manifold factors onto the group of
orientation-preserving symmetries of a dodecahedron. This was enough
to show that that manifold could not be the three-sphere.

Now Poincaré gave no indication as to how he obtained this remarkable
example: remarkable indeed, because it is the *only* known homology
three-sphere with finite fundamental group, besides the three-
sphere itself. In fact, the fundamental group G is a central ex-
tension of $\mathbb{Z}/2\,\mathbb{Z}$ by the alternating group on five elements A_5 (which
is also the group of orientation-preserving symmetries of the dodecahedron). This group G has, therefore, 120 elements and is a
perfect group, in as much as its abelianization, being the first
homology group of the manifold, is trivial.

Ten years prior to this discovery of Poincaré, Klein published his
book "Vorlesungen über das Ikosaeder und die Auflösung der Glei-
chungen vom fünften Grade" (Leipzig, 1884). In it, Klein considers
the group SO(3) of orientation-preserving isometries of three-
dimensional euclidean space fixing the origin, and classifies the
subgroups of finite order, showing that they consist of the groups
and subgroups of symmetry of the platonic solids and of the n-
bipyramid. Thus A_5 is a subgroup of SO(3) which divides SO(3) in
right classes SO(3)/A_5. Via the stereographic projection, SO(3) can
be considered as a subgroup of the group of homographies of the
complex projective line. This group is double covered by the linear
group SL(2,$\mathbb{C}$) acting in $\mathbb{C}^2$. Klein considers the quotient $\mathbb{C}^2/\Gamma$,
where Γ is a finite subgroup of SL(2,$\mathbb{C}$) and shows that $\mathbb{C}^2/\Gamma$ is an
algebraic variety which can be represented as a hypersurface of $\mathbb{C}^3$.
This hypersurface has no singularity except at the origin $o\in\mathbb{C}^3$
and a neighbourhood boundary of it can be identified with S^3/Γ. If
Γ double-covers $A_5\subset$SO(3), S^3/Γ (or SO(3)/A_5) is a three-manifold.
It turns out that this manifold is the Poincaré homology sphere.
Thus, as L. Siebenmann pointed out to me, Klein must have been in
possession of the Poincaré homology sphere years beforehand, and in
a different and more instructive way than Poincaré, but perhaps not
knowing its important properties (compare [S1]).

This foregoing description shows immediately that the Poincaré
homology sphere is the space of positions of a dodecahedron
inscribed in the sphere. Then one might ask why we should not just
study SO(3)/A_5 instead of that space of positions. My contention is
the following: firstly, I feel that the space of positions is more
intuitive than the cold definition SO(3)/A_5 . Secondly, it has
advantages, the least of which is that it allows us to give examples of three-manifolds in a very simple way. Besides this, we
would do well to consider the following. Take a fixed diameter of
the sphere and define an action of the one-dimensional sphere S^1
in the Poincaré homology sphere by rotating positions around the
diameter. We note that if we take a position of the dodecahedron
having the diameter as an axis of two, three or five-fold symmetry,
the dodecahedron comes back to its original position at one half,
one third, or one fifth of a turn, respectively. For all other
positions, you need a complete turn to come back. Therefore, our
space of positions is the disjoint union of circles, but three of
these circles are *exceptional*. This defines a Seifert-manifold

structure in the Poincaré homology sphere with exceptional fibers
of orders two, three and five. Clearly, if the diameter is changed,
we obtain other Seifert-manifold structures but all of them are
conjugate to each other.

This is the type of topic we deal with in this book, only that
instead of restricting our attention to the dodecahedron, we also
consider the remaining platonic solids, and the euclidean and hyperbolic tessellations for which analogous constructions of three-
manifolds can be developed in a similar way. At this stage one might
also ask what can be considered new here. In fact, there is nothing
new except the point of view. What I had in mind in writing this
book was to use these constructions as a "pretext" for talking about
three-manifolds and teaching geometrical intuition, which is crucial
in forming our students to be able to make new discoveries in
mathematics. Thus these constructions serve as a vehicle to introduce several concepts about three-manifolds.

The graduate student wishing to work in low-dimensional topology will
find here a source of geometrical insight. Professionals will see
what they already know clothed in a different garb, and they might
use the book as a source of teaching ideas for their low-dimensional
topology seminars, to prepare independent study projects for their
students, or as the basis of a reading course.

After a summary glance over this book, one soon sees it is in some
ways an "unorthodox" book: on the one hand, there are no explicit
statements of theorems but, on the other hand, there are many
geometrical descriptions illustrated with diagrams. It becomes
clear that this is not a textbook in the usual sense of the word,
but that it will be helpful to have it available when reading related low-dimensional material.

The fundamentals for many constructions in the book can be found
in other sources, in particular [GP] and parts of [RS], [R] and
[ST]. I assume some knowledge of projective geometry ([VY], [Bl],
[Cox2], [Le]), hyperbolic geometry ([BE]), and algebraic topology
([ST] or [Ma] and [GH]).

The book is the result of a series of lectures first given at
Zaragoza and then at Oaxtepec (Mexico) in 1982, within the

"VI Escuela Latinoamericana de Matématicas". A *condition sine qua non* was to deliver a manuscript for the use of those attending it. It was first written in Spanish and distributed as a set of notes to the participants. Later I decided to give the book a more modern flavor. I was influenced by Thurston's results which showed the importance of his eight geometries and the use of orbifolds. Now the manifolds of positions of tessellations, studied in this book, illustrate, but do not exhaust, three of the eight geometries; namely the universal covering spaces of the unit tangent bundles of the euclidean, spherical and hyperbolic planes. Consequently, I have added an appendix on orbifolds and use the language of orbifolds throughout the book.

Following the suggestions of two of my colleagues at Zaragoza, I refrained from increasing the length of the book, and decided ultimately to avoid including material about Dehn-surgery, Heegaard diagrams, branched coverings, etc. All this material can be found in Rolfsen's book [R]. Inevitably then, the present book is somewhat monographic and I readily admit that the repetition of the same idea in Chapters Two, Three and Five might be considered a little tedious. However, that idea is treated in different ways, using different techniques.

In short, technically the book deals with the unit tangent bundle of the two-dimensional orbifolds. This topic of orbifolds consti-tutes the appendix to Chapter Two. Chapter One studies unit tangent bundles of two-manifolds. Chapters Two, Three and Five are devoted to unit tangent bundles of euclidean, spherical and hyperbolic orbifolds, while Chapter Four studies Seifert-manifolds. The appendix to Chapter Five is a draft for an alternative description of the hyperbolic plane, including ideas which hopefully can be seen as fitting into the theme of the book.

Finally, I must solicit the reader's indulgence for the biblio-graphical lacunae which he or she will readily notice, as my purpose in writing this book was not to attempt a source-book for scholars but rather, as I said earlier, to provide for graduate students a source of geometrical intuition in low-dimensional topology

JMM
Berkeley, May 13, 1987

Acknowledgements

I am deeply indebted to my colleague María Teresa Lozano who gave
me good advice and read the book critically. Francis Bonahon and
Alexis Marin pointed out some mistakes and suggested corrections.
W. Dumbar, K. Luttinger, P. Ney de Souza, L. Rudolph, C. Safont and
W. Whitten read the book very carefully, improved my English translation and provided very interesting comments. I want to record here
my warmest thanks to all of them. Thanks are due to H.M. Hilden for
an enlightening conversation, to L. Siebenmann for his helpful advice and to J. Besteiro, from the Department of Crystallography of
Zaragoza, who drew some of the ornaments at the end of the chapters
from their Mudéjar originals. Thanks are due also to J.S. Birman for
her encouragement, and to José Adem who originally invited me to
deliver the lectures at Oaxtepec (Mexico). Finally I would like to
thank Torlach Delargy for his assistance in the translation of some
of the more difficult sections, and to Pablo del Val for his help
in the proofreading of the manuscript. I would like to thank the
staff of Springer-Verlag for the pleasant cooperation we had during
the various stages of development of the book, and also Mrs. Jebram
for her efficient typing of the final copy of the manuscript.

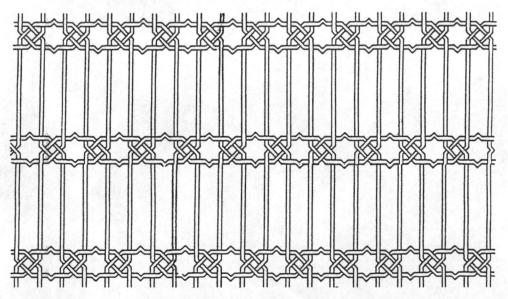

Acknowledgements

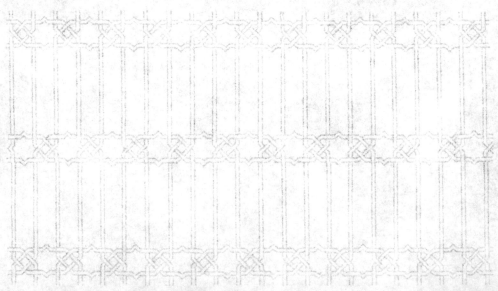

Table of Contents

Chapter One: S¹-Bundles Over Surfaces

> "Cap. LXXVII, Donde se cuentan mil zarandajas, tan
> impertinentes, como necesarias al verdadero
> entendimiento de esta grande Historia."
>
> *"Chapter LXXVII. In which a Thousand Trifles are*
> *recounted, as nonsensical as they are necessary to*
> *the True Understanding of this great History."*
> Cervantes, Don Quixote

A *closed surface* is a compact, connected 2-manifold without
boundary. From the tangent bundle TX of a closed surface X we can
construct the *spherical* (or *unit*) *tangent bundle* of X, denoted
ST(X), as the subbundle of TX consisting of vectors of norm 1 (see
[GP], page 55). The fiber of ST(X) is the 1-sphere S^1, and thus
ST(X) is a closed 3-manifold, which always has a canonical orientation, even when the base X of the bundle is non-orientable (see
[GP], pp. 76 and 106). These S^1-bundles were known by the suggestive
name of "bundles of oriented line elements" (see [ST]). The "bundles
of unoriented line elements" of X, denoted by PT(X), are obtained
from ST(X) by identifying the vectors (x, v) and (x, -v), for every
(x, v) ∈ ST(X). The S^1-bundle PT(X), which is also called *projective*
tangent bundle, is a canonically oriented, closed 3-manifold and the
natural map ST(X)→PT(X) is a 2-fold covering.

In this chapter we describe these manifolds and also their generali-
zations, namely the S^1-bundles over surfaces. It is very convenient
to treat this topic here since the whole book deals with a generalization of it. As we said in the preface the manifolds of tessellations, the subject of this book, are nothing but the spherical
tangent bundles of 2-dimensional orbifolds (a generalization of
surfaces).

1.1 The spherical tangent bundle of the 2-sphere S^2

We start by describing $ST(S^2)$, the spherical tangent bundle of the unit 2-sphere

$$S^2 := \{\vec{x} \in \mathbb{R}^3 : |\vec{x}| = 1\}.$$

This is very easy to understand and helps to give the flavor of the situation.

Each element of $ST(S^2)$ is a point in S^2 together with a tangent unit vector at this point (a *pointer*, compare [HC], p. 69). If we fix some *base pointer* b, any other pointer b' *uniquely* defines a rotation of $\mathbb{R}^3$ sending b to b', as is shown in Fig. 1. This proves that $ST(S^2)$ is just the Lie group $SO(3)$ of orientation-preserving isometries of the euclidean space E^3 fixing the origin. It is well known that $SO(3)$ consists of rotations.

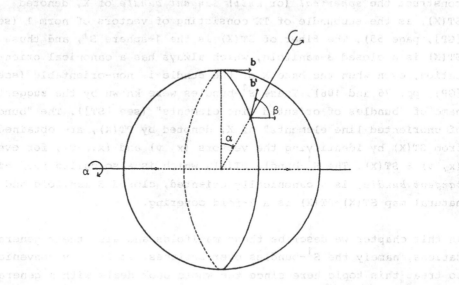

Fig. 1. Rotation defined by the pointer b'

Following Seifert and Threlfall, we now give a different description of $SO(3)$, which will be tackled again in 3.6 with the help of quaternions. Take the ball $B^3 \subset \mathbb{R}^3$ of radius π. To each point x of

B^3 we can associate the rotation of axis ox and angle $|x|$ (to
o we associate the identity). Since antipodal points of ∂B^3 rotate
$\mathbb{R}^3$ around the same axis and with angle π, they define the same
rotation of $\mathbb{R}^3$ and, therefore, we identify them. This shows that
$ST(S^2)$ ($\cong SO(3)$) is the 3-manifold obtained from a ball B^3 by
identifying antipodal points of its boundary. This manifold is the
real projective space $\mathbb{R}P^3$ (this is left as an exercise to the
reader).

Thus we see that $ST(S^2)$ is $\mathbb{R}P^3$. But there is another interpretation
of $ST(S^2)$ which is of great interest. An element of $ST(S^2)$, i.e.
a pointer v based in some $x \in S^2$, defines a family of curves passing
through it with velocity v. Among them select a geodesic, i.e. a
great circle with constant velocity v, and assume that x travels
along it. In time t=1, say, x reaches the place y, and clearly the
pair (x, y) represents v, because there is a unique geodesic passing
through x and reaching y in time t=1. The point (x, y) belongs to
$S^2 \times S^2$ and we see that the D^2-bundle of S^2, consisting of vectors
of $T(S^2)$ of norm ≤ 1, is a tubular neighbourhood of the diagonal

$$\Delta = \{(x, y) \in S^2 \times S^2 \mid x=y\}$$

of $S^2 \times S^2$ (see Fig. 2 and [GP], page 76, for more details).

Therefore $ST(S^2)$ ($\cong \mathbb{R}P^3$) is the spherical normal bundle $SN(\Delta)$ of Δ
in $S^2 \times S^2$. This has a natural orientation because the local
orientation around $(x, x) \in \Delta$ is given by the product $E_x \times E_x$,
where the orientation selected for E_x is immaterial (see Fig. 2,
where $E_x = \{y \in S^2 : x$ can be joined to y by a geodesic of length
$\leq 1\}$).

Notice that $SN(S^2 \times o)$ in $S^2 \times S^2$ is the trivial S^1-bundle of S^2,
i.e. the product bundle $S^2 \times S^1$. But $\mathbb{R}P^3$ and $S^2 \times S^1$ are different
(their fundamental groups are $\mathbb{Z}_2$ and $\mathbb{Z}$, respectively) thus we see
that it is possible to embed S^2 in $S^2 \times S^2$ in at least two different
ways, namely Δ and $S^2 \times o$. An algebraic proof of this same fact
involves looking at the intersection form

$$\cdot : H_2(S^2 \times S^2; \mathbb{Z}) \times H_2(S^2 \times S^2; \mathbb{Z}) \to \mathbb{Z}$$

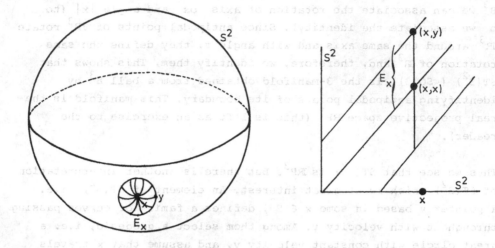

Fig. 2. The fiber over x ∈ S²

(see [ST]):

·	a	b
a	0	1
b	1	0

where a, b are represented by $S^2 \times o$ and $o \times S^2$. The inequality

$$\Delta \cdot \Delta = (a+b) \cdot (a+b) = 2 \neq 0 = a \cdot a$$

also shows that the embeddings Δ and $S^2 \times o$ are different.

Later in this chapter, we will show that the self-intersection number of $i(S^2)$ in $S^2 \times S^2$, for any embedding

$$i: S^2 \hookrightarrow S^2 \times S^2,$$

determines $SN(i(S^2))$. Thus, the two methods for checking that the two embeddings are different are, after all, the same. This self-intersection number is called the *Euler number* of the S^1-bundle $SN(i(S^2))$, and it classifies the bundle. The name "Euler number" is given because, for the diagonal embedding Δ, the Euler number $\Delta \cdot \Delta$ is the Euler characteristic $\chi = 2$ of the 2-sphere.

In section 1.2 we will introduce the Euler number of S^1-bundles of oriented closed surfaces.

Exercise 1. *Is it true that* $ST(S^3)$ *coincides with* $SO(4)$? *Does* $ST(S^1)$ *coincide with* $SO(2)$?

Exercise 2. *Find embeddings of* S^2 *in* $S^2 \times S^2$ *with self-intersection number* $2m$, $m > 1$.

1.2 The S^1-bundles over oriented closed surfaces

We now describe the S^1-bundles of an *oriented, closed surface of genus g*, F_g. This surface is a sphere with g handles. The case g=2 is depicted in Fig. 3.

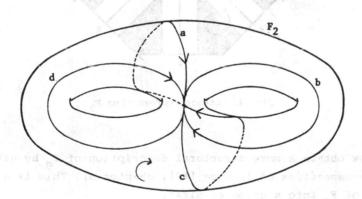

Fig. 3. Oriented surface of genus g = 2

To work more efficiently we cut F_g along 2g curves, like the curves a, b, c, d of F_2 represented in Fig. 3. The result of cutting F_g open along them is a 4g-sided polygon. The octagon corresponding to g=2 is depicted in Fig. 4. The surface F_g is this 4g-polygon with pairs of sides identified. This is a more abstract picture of F_g than the one of Fig. 3, but it has the advantage of lying in a plane.

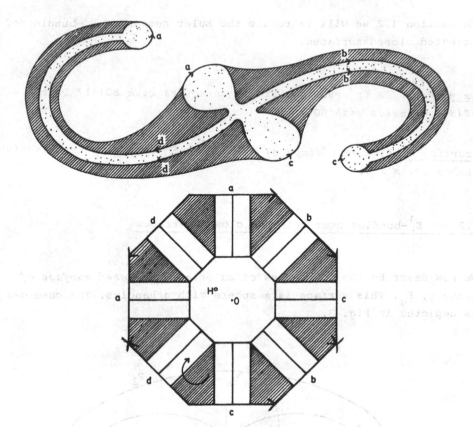

Fig. 4. Octagon representing F_2

We can now obtain a more structural description of F_g by giving a
handle decomposition of it (see [RS], chapter 6). This is a decomposition of F_g into a union of disks,

$$F_g = H^0 \cup 2g\ H^1 \cup H^2 ,$$

that we can understand "dynamically" as follows. The central point
o of the polygon grows (in two dimensions) to give a smaller polygon
called H^0 (the "0-handle"), where the initial "condensation point"
is called the *core* of the 0-handle H^0 (see Fig. 4). To H^0 we attach
the *cores of the 2g 1-handles* (these are *1-dimensional* disks, and
that is why the corresponding handles are referred to as *1-handles*).
These 2g cores are the arcs shown in Fig. 4. Then we make these
cores grow (in two dimensions) to produce the 2g disks called
1-handles (Fig. 5). To

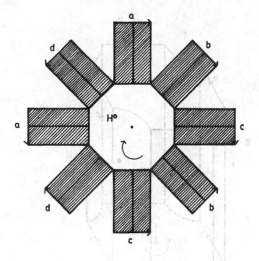

Fig. 5. 0- and 1-handles of F_2

$$H^0 \cup 2g \ H^1$$

we finally attach one 2-handle H^2. The core of this 2-handle, being a 2-dimensional disk, coincides with the 2-handle. In Fig. 4 H^2 is the union of the sectors containing the corners of the 4g-polygon.

We are now in a position to describe the S^1-bundles of F_g. Assume M^3 is an *oriented* S^1-bundle over F_g. Then the restriction of M to each one of the handles of F_g is a trivial bundle (a solid torus) because each handle is a disk (see [St], page 53). This is the reason for decomposing F_g into handles. Thus M^3 is just

$$(H^0 \times S^1) \cup 2g(H^1 \times S^1) \cup (H^2 \times S^1) \ .$$

The oriented solid torus $H^0 \times S^1$, for g = 1, appears in Fig. 6. We attach to it the 2g solid tori $H^1 \times S^1$. This is equivalent to identifying in pairs the 4g annuli of $\partial(H^0 \times S^1)$, which in Fig. 6 are marked II, III. Observe that these identifications are *essentially unique* because M is *oriented* (see Fig. 6). Thus we see that, because M is oriented, the part of M lying over $H^0 \cup 2g \ H^1$ is a product. In other words, the orientability character of the bundle M is determined by the restriction of M to the "1-skeleton" of F_g (i.e. the result of shrinking $H^0 \cup 2g \ H^1$ by shrinking first 2g H^1

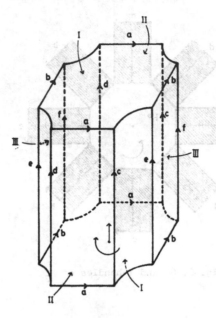

Fig. 6. $H^0 \times S^1 \cup 2g(H^1 \times S^1)$, $g = 1$

to the cores of $2g \ H^1$ and then H^0 to the core of H^0). The bundle over the "2-skeleton" (i.e. the final bundle) is obtained by identifying the boundaries of

$$W := (H^0 \cup 2g \ H^1) \times S^1$$

and $H^2 \times S^1$; and it is here that the Euler number appears.

The manifold

$$(H^0 \cup 2g \ H^1) \times S^1$$

is oriented by the product of the orientations of $H^0 \cup 2g \ H^1$ and S^1. Its oriented boundary ∂W is a torus in which we fix the oriented meridian-longitude basis

$$(m, \ell) := (\partial(H^0 \cup 2g \ H^1) \times x, \ y \times S^1) \ ,$$

where x is some point of S^1 and y is some point of $\partial(H^0 \cup 2g \ H^1)$. Thus

$$m \cdot \ell = 1 \ ,$$

on ∂W (see Fig. 7). Now we identify ∂W with $\partial(H^2 \times S^1)$ so that the result will be an S^1-bundle over F_g. This identification process is equivalent to the following two operations. First, fiber ∂W by the 1-spheres C_z which are the images of $\partial H^2 \times z$, for every $z \in S^1$ (the only condition the C_z's must satisfy to get an S^1-bundle is that they cut each fiber of ∂W in precisely one point). Finally, collapse each curve C_z to one point (this is the same as pasting the disk $H^2 \times z$ to the circle $C_z = \partial H^2 \times z$).

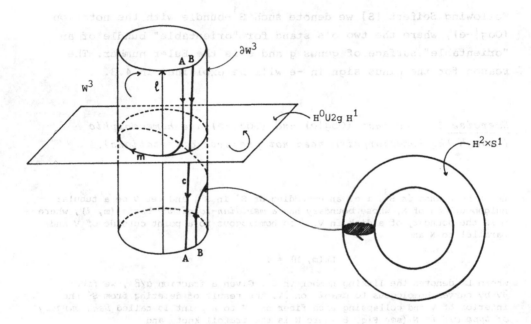

Fig. 7. W identified with $H^2 \times S^1$; e = -2

We see that the final S^1-bundle M depends *only* on the fibration of ∂W by the curves C_z. This fibration is characterized by the algebraic intersection of m with C (= element of $H_1(\partial W; \mathbb{Z})$ determined by any C_z), where C_z is oriented so as to have

$$C \cdot \ell = 1 ,$$

on ∂W (Fig. 7). *The integer*

$$e := m \cdot C ,$$

on ∂W, *determines the S^1-bundle up to orientation-preserving equivalence, and is called the Euler number of the S^1-bundle.* It satisfies

$$C = m + e\ell$$

on ∂W, and in some sense measures the obstruction to having a section of the S^1-bundle (in fact, e = 0 if and only if that section exists).

Following Seifert [S] we denote such S^1-bundle with the notation (Oog|-e), where the two o's stand for "orientable" bundle of an "orientable" surface of genus g and e is the Euler number. The reason for the minus sign in -e will be explained in 4.3.

Exercise 1. Show that (Oog|e) *and* (Oog|-e) *are homeomorphic (though the homeomorphism need not preserve orientations).*

Let N be a *knot in* S^3, i.e. an embedding of S^1 in S^3, and let V be a tubular neighbourhood of N, whose boundary has a *meridian-longitude* basis (m, ℓ), where m is the boundary of a disk in V, ℓ is homologous to a point outside of V and parallel to N and

$$Lk(m, N) = 1$$

where Lk denotes the linking number in S^3. Given a fraction α/β , we fiber ∂V by curves homologous to $\alpha m + \beta \ell$ on ∂V. The result of deleting from S^3 the interior of V and collapsing each fiber of ∂V to a point is called *Dehn surgery of type α/β in* N (see Fig. 8 where N is the trefoil knot and

$$\alpha/\beta = 0.$$

Our convention for figures is that S^3 is supposed to be oriented, and that in the pictures, this orientation is given by a right-handed screw). For more details on Dehn surgery see the beautiful book [R].

Exercise 2. Show that (OoO|-e) *is the result of Dehn surgery e in the trivial knot.*

Exercise 3. Show that 1/β *Dehn surgery in the trivial knot gives back* S^3.

Hint. (Hempel's trick) Cut S^3-int V open along a disk bounding N. Twist one side a number β of full twists and paste the sides back together.

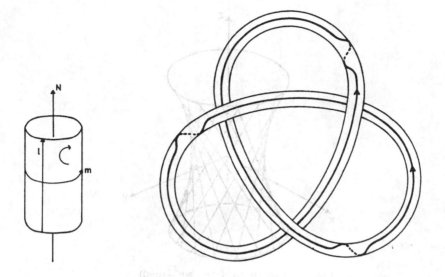

Fig. 8. Dehn surgery 0 in the trefoil knot

The manifold $L(\alpha, \beta)$ obtained by $-\alpha/\beta$ Dehn surgery in the trivial knot (g.c.d.(α, β)=1) is called a *lens space* (see [ST]). Thus (Oo0|b) is L(b, 1).

1.3 The Euler number of $ST(S^2)$

The Euler number of $S^2 \times S^1$ is zero. We now compute that number for $ST(S^2)$ ($\cong \mathbb{R}P^3$) and we expect to obtain 2 (see 1.1).

Using homogeneous coordinates $(x : y : z : u)$ in $\mathbb{R}P^3$, the equation

$$x^2 + y^2 - z^2 - u^2 = 0$$

represents the ruled hyperboloid, in the affine space $\mathbb{R}P^3 - (u = 0)$, depicted in Fig. 9. This surface decomposes $\mathbb{R}P^3$ into two components. The one containing the z-axis is a solid torus. But what is the complementary region? If we change coordinates by means of the formulae

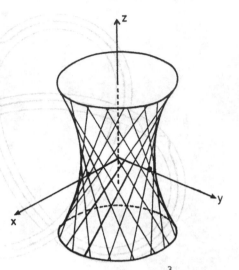

Fig. 9. Hyperboloid in $\mathbb{RP}^3-(u=0)$

$$x + z = x', \quad x - z = z'$$
$$u + y = u', \quad u - y = y'$$

the new equation is

$$x'z' = y'u' \quad ,$$

which in $\mathbb{RP}^3 - (u'=0)$ is the ruled paraboloid shown in Fig. 10.
The reflection in the x'-axis, interchanging both sides of the paraboloid, shows that $\mathbb{RP}^3$ is the union of two solid tori V and W which
are pasted along their common boundary. The meridians C and m of V
and W (also permuted by the reflection) are the ascending and
descending parabolas cutting themselves in the saddle point
(0:0:0:1) and at the infinite point (0:0:1:0). Thus the Euler
number of $\mathbb{RP}^3$ is

$$e = m \cdot C$$

(on ∂W), assuming

$$m \cdot \ell = 1$$

and

$$C \cdot \ell = 1$$

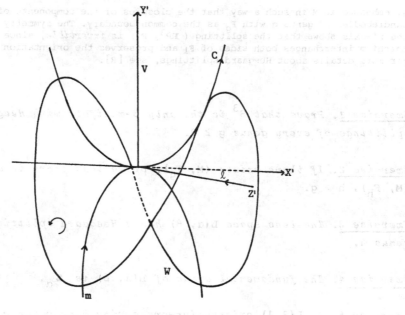

Fig. 10. $\mathbb{R}P^3$ is the union of two solid tori

(Fig. 10). *Thus, e = -2; hence $\mathbb{R}P^3$ is the lens space*

$$(Oo0|2) = L(2, 1) \ .$$

Since the antipodal map in $\mathbb{R}P^3$,

$$(x:y:z:u) \mapsto (-x: -y: -z: u) \quad ,$$

reverses the orientation of $\mathbb{R}P^3$, we see that $\mathbb{R}P^3$ is a *symmetric manifold* in the sense that it admits an orientation-reversing automorphism. Thus

$$\mathbb{R}P^3 = (Oo0|2)$$

is orientation-preserving equivalent to

$$- \mathbb{R}P^3 = (Oo0|-2) \ .$$

The pair $(\mathbb{R}P^3, F_1)$, where F_1 is the hyperbolic paraboloid, is an example of a *Heegaard splitting*. A Heegaard splitting of genus g is a pair (M, F_g) composed of a closed, orientable 3-manifold M and a closed, orientable surface F_g of genus g, embedded in M in such a way that the closures of the components of $M-F_g$ are handlebodies of genus g with F_g as the common boundary. The symmetry around the x'-axis shows that the splitting ($\mathbb{R}P^3$, F_1) is *invertible*, since that homeomorphism interchanges both sides of F_1 and preserves the orientation of $\mathbb{R}P^3$. For more details about Heegaard splittings, see [R].

Exercise 1. *Prove that* S^3 *is the only 3-manifold with Heegaard splittings of every genus* g $\geq$ 0.

Exercise 2. *If there exists* (M, F_g), *prove that there is also* (M, F_h), h > g.

Exercise 3. *The lens space* L(α, β) *has a Heegaard splitting of genus* 1.

Exercise 4. *The fundamental group of* L(α, β) *is* $\mathbb{Z}_\alpha$.

Exercise 5. *Is* L(3,1) *orientation-preserving homeomorphic to* -L(3,1), *or, in other words, is* L(3,1) *symmetric? (see* [ST], *page 291).*

1.4 The Euler number as a self-intersection number

So far, we have computed the Euler number of $S^2 \times S^1$ and of

$$ST(S^2) \cong \mathbb{R}P^3 \quad .$$

We want also to compute that number for $ST(F_g)$. To do this we want to interpret the Euler number in a way which is very useful in 4-dimensional topology. In the following, Lk is the linking number in S^3 (see [ST]).

Let ξ be a D^2-bundle over F_g with orientable total space W^4. Then, the Euler number of the S^1-bundle ∂W^4 is $F_g \cdot F_g$. In fact let P^2 be a section of the S^1-bundle ∂W^4 restricted to

$$H^0 \cup 2g \, H^1 = F_g - H^2 \quad .$$

We attach to P a disk E^2 in $H^2 \times D^2$ bounded by $\partial P = m$. Thus we obtain a closed oriented 2-manifold $\hat{F}_g$ which is homologous to F_g in W^4. Then

$$F_g \cdot F_g = F_g \cdot \hat{F}_g = (H^2 \times 0) \cdot E^2$$

in $H^2 \times D^2$ (with this product orientation). But

$$(H^2 \times 0) \cdot E^2 = Lk(\partial(H^2 \times 0), \partial E^2)$$

in $\partial(H^2 \times D^2)$, where the orientation here is such that

$$Lk(\partial H^2 \times 0, 0 \times \partial D^2) = 1 .$$

But

$$\partial E^2 = -m = -C + e\ell ,$$

where C bounds a disk in the solid torus $H^2 \times \partial D^2$, e is the Euler number of ∂W^4 and $\ell = 0 \times \partial D^2$. Thus

$$Lk(\partial H^2 \times 0, \partial E^2) = Lk(\partial H^2 \times 0, -m)$$

$$= Lk(\partial H^2 \times 0, -C + e\ell)$$

$$= Lk(\partial H^2 \times 0, e\ell)$$

$$= e\, Lk(\partial H^2 \times 0, 0 \times \partial D^2)$$

$$= e$$

(Fig. 11). Now we will show that the *spherical tangent bundle of an orientable surface F_g is* $(0\ o\ g\ |\ -\chi(F_g))$, i.e. *its Euler number coincides with the Euler characteristic of F_g*. The starting point is [GP], page 76, where it is shown that $ST(F_g)$ coincides with the spherical normal bundle $SN(\Delta)$, where Δ is the diagonal of $F_g \times F_g$ (alternatively: adapt the proof of 1.1 to F_g, using a Riemannian metric in F_g, [DoC]). Thus the Euler number of $ST(F_g)$ is then $\Delta \cdot \Delta$ in $F_g \times F_g$. Now we claim that $\Delta \cdot \Delta$ is $\chi(F_g)$. This can be shown directly as indicated in the following exercise.

Exercise 1. Let F_g be as in Fig. 12, i.e. there is a "symplectic" basis in F_g, such that

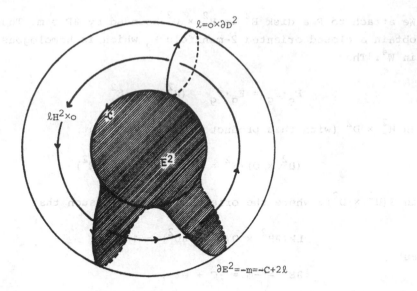

$\ell = 0 \times \partial D^2$

$\ell H^2 \times 0$

E^2

$\partial E^2 = -m = -C + 2\ell$

Fig. 11. Construction of E^2 in $\partial(H^2 \times D^2) = \mathbb{R}^3 + \infty$; e = 2

$$m_i \cdot \ell_i = 1 \ ,$$

$$m_i \cdot m_j = \ell_i \cdot \ell_j = m_i \cdot \ell_j = 0 \ , \quad i \neq j.$$

Show that
$$F_g \times 0, \ 0 \times F_g, \ m_i \times m_j, \ m_i \times \ell_j, \ \ell_i \times m_j, \ \ell_i \times \ell_j$$
form a basis for $H_2(F_g \times F_g; \ \mathbb{Z})$ *with* $2 + 4g^2$ *elements. Find the intersection form, and compute* $\Delta \cdot \Delta$ *noting that*

$$\Delta = (F_g \times 0) + (0 \times F_g) + \sum_{i=1}^{g} m_i \times \ell_i + \sum_{i=1}^{g} \ell_i \times m_i \ .$$

However we give a different proof which is more conceptual. To compute $\Delta \cdot \Delta$ we move Δ slightly and smoothly to obtain Δ' transverse to Δ; i.e. $\Delta \cap \Delta'$ consists of a finite number of points where the intersection occurs like the two coordinate planes of $\mathbb{R}^2 \times \mathbb{R}^2$ at the origin. Then

$$\Delta \cdot \Delta = \Delta \cdot \Delta'$$

and we only need to understand the signs at the points of $\Delta \cap \Delta'$ (see [ST]). All of this has a different interpretation. If we assume that F_g has a Riemannian metric (so that nearby points are connected by unic geodesics), the point $(x, y) \in \Delta'$ can be thought

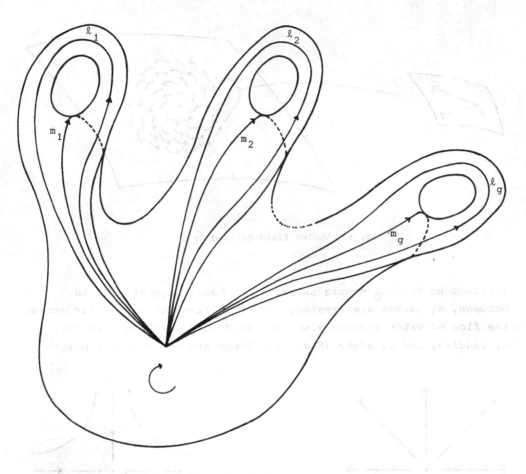

Fig. 12. Surface F_g

of as a little tangent vector to F_g based at x and "ending" in y
(see 1.1 and Fig. 13). This creates a vector field, and since the
motion of Δ to obtain Δ' is smooth, this vector field is also
smooth, and the singular points correspond to the points
(x, x) ∈ Δ', i.e. to the points of Δ∩Δ'. The fact that Δ and Δ'
intersect transversally can be considered as a definition for the
field to have only *isolated* (or generic) singular points.

The converse is also true, and any such vector field defines Δ'
transverse to Δ in $F_g \times F_g$. We can take a particularly beautiful
vector field on F_g as follows. First triangulate F_g with "smooth"
triangles so that if two triangles meet at all, they do so in a
common face. Sink the centers of the triangles and raise the

Fig. 13. Vector field defined by Δ'

vertices so that α_0 mountains arise, α_2 basins appear and, in between, α_1 passes are created. If now it rains over this landscape, the flow of water induces a vector field on F_g with α_0 *sources*, α_1 *saddles*, and α_2 *sinks* (Fig. 14). These are the singular points

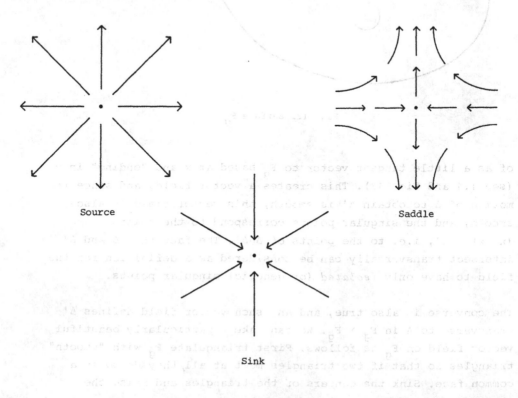

Source   Saddle

Sink

Fig. 14. Some singular points

of the vector field, and if we create Δ' using it, Δ∩Δ' consists
precisely of those points. We must understand the sign that they
contribute to the intersection number Δ·Δ'. This is shown in Fig.
15. We take a little circle C around the singular point x. We consider the solid torus $C \times D^2$ consisting of fibers of

$$T(F_g) = N(\Delta)$$

lying over C. The solid torus is naturally oriented, and we select
a trivialization of N(Δ) restricted to the disk in F_g with center x
and boundary C. The sign(x) coincides with the linking number
Lk(C, C'), where C' is Δ'∩(C × D²) (see [RS], page 69). This linking
number is computed for a source and a saddle; for a sink it is
left as an exercise to see that it is +1.

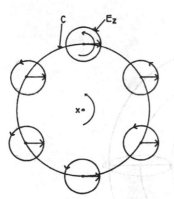

Orientation and
trivialization

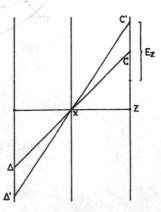

Source

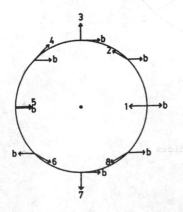

Saddle

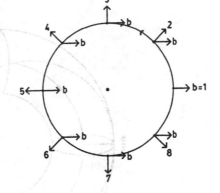

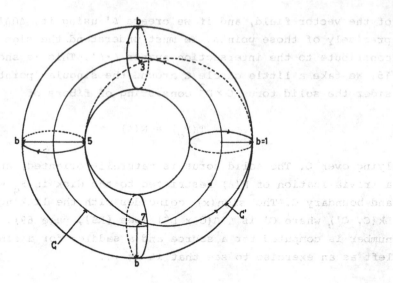

20

sign (x) = Lk (C,C') = 1

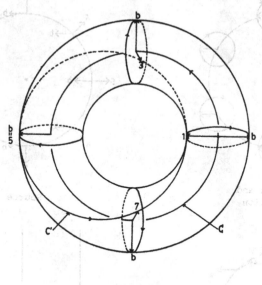

sign(x) = Lk (C,C') = -1

Fig. 15. Computing indices

Remark.- In Fig. 15 we have computed the *indices* of some particular
singular points. In [GP], page 133, a general definition of index
of a singular point is given. These indices coincide with the sign
of the points of $\Delta \cap \Delta'$ corresponding to the singular points. Thus
the sum of indices of *any* vector field of F_g coincides with

$$\Delta \cdot \Delta' = \Delta \cdot \Delta ,$$

i.e. with $\chi(F_g)$. This is the theorem of Poincaré-Hopf ([GP], page
134).

These beautiful results explain the name Euler number and justify
the fact, already proved, that $ST(S^2)$ has Euler number 2.

*Exercise 2. Compute the Euler number of the spherical normal
bundle of a surface F_g standardly embedded in S^4 (i.e.*
$F_g \subset \mathbb{R}^3 \subset \mathbb{R}^4 \subset S^4$).

1.5 The Hopf fibration

The complex projective plane $\mathbb{C}P^2$ is the result of compactifying
$\mathbb{C}^2$ by adding an ideal point at infinity to each complex line
passing through the origin of $\mathbb{C}^2$. Each of these complex lines becomes a

$$\mathbb{C}P^1 \cong S^2 .$$

Thus we have a "bundle" of complex projective lines, whose infinite
points also form a $\mathbb{C}P^1$ which is called the "projective line at
infinity".

The 3-sphere

$$S^3 \subset \mathbb{C}^2 \subset \mathbb{C}P^2$$

separates $\mathbb{C}P^2$ into two components whose closures are D^4 and a

D^2-bundle whose base space is the projective line at infinity. Thus S^3 is an S^1-bundle with base S^2 (*Hopf fibration*) whose Euler number is

$$\mathbb{CP}^1 \cdot \mathbb{CP}^1$$

in $\mathbb{CP}^2$, i.e. 1, for the orientation of S^3 as boundary of the D^2-bundle considered, and therefore -1 for its standard orientation as boundary of D^4.

To illustrate the projection of the bundle, we project S^3 stereographically onto $\mathbb{R}^3$ (Fig. 16). The points (1, 0), (i, 0), (0, 1), (0, i) are O, e_1, e_2, e_3 respectively.

The fiber of S^3 passing through (u, v) is the set of pairs $e^{it}(u, v)$, $0 \leq t < 2\pi$; hence it is contained on the torus

$$\{(z_1, z_2) \in S^3 : z_1\bar{z}_1 = u\bar{u}, \ z_2\bar{z}_2 = v\bar{v}\}$$

(Fig. 17). This torus degenerates into the circles E, F when $u\bar{u} = 1$ or $v\bar{v} = 1$. The bundle of tori is depicted in Fig. 18, and the way in which each torus is fibered, in Fig. 17.

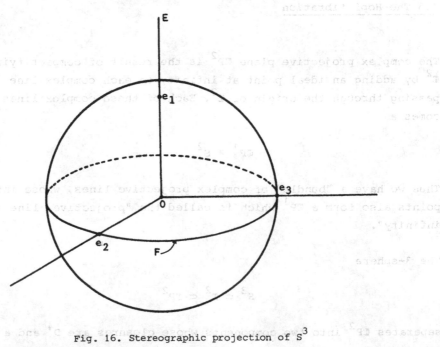

Fig. 16. Stereographic projection of S^3

23

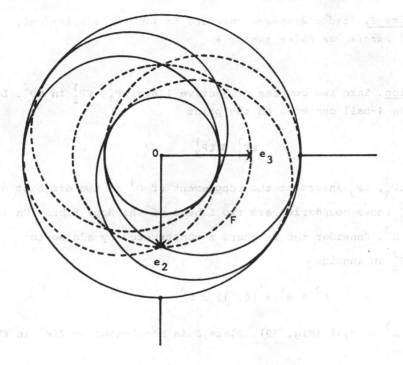

Fig. 17. Fibration of a torus

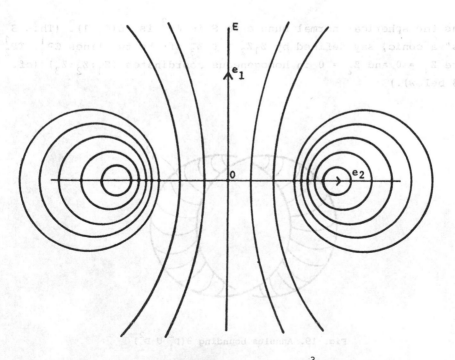

Fig. 18. Bundle of tori in S^3

Exercise 1. Find a 2-sphere embedded in $\mathbb{C}P^2$ *whose spherical, normal bundle has Euler number* 4.

<u>Solution</u>. Take two complex projective lines $\mathbb{C}P^1_1$, $\mathbb{C}P^1_2$ in $\mathbb{C}P^2$. Let D^4 be a 4-ball centered on the point

$$\mathbb{C}P^1_1 \cap \mathbb{C}P^1_2 \ .$$

Then $\mathbb{C}P^1_1$, $\mathbb{C}P^1_2$ intersect the complement of D^4 in two disjoint discs D^2_1, D^2_2 whose boundaries are two fibers of the Hopf fibration in $\partial D^4 = S^3$. Consider the 2-sphere S constructed by adding to $D^2_1 \cup D^2_2$ an annulus

$$A^2 = S^1 \times [0, 1] \subset S^3$$

along $S^1 \times \{0,1\}$ (Fig. 19). Since S is homologous to $2\mathbb{C}P^1$ in $\mathbb{C}P^2$ we have

$$S \cdot S = (2\mathbb{C}P^1)^2 = 4 \ .$$

Thus the spherical normal bundle to S in $\mathbb{C}P^2$ is -L(4, 1). (This S "is" a conic, say defined by $Z_1 Z_2 = \varepsilon\ Z^2_3$ if the two lines $\mathbb{C}P^1_1$, $\mathbb{C}P^1_2$ were $Z_1 = 0$ and $Z_2 = 0$ in homogeneous coordinates $(Z_1 : Z_2 : Z_3)$ (cf. 1.8 below).)

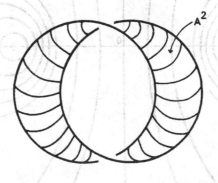

Fig. 19. Annulus bounding $\partial(D^2_1 \cup D^2_2)$

1.6 Description of non-orientable surfaces

We will describe the non-orientable surfaces before studying their S^1-fiber bundles.

A non-orientable surface of genus k, N_k, is the result of deleting the interior of k disks from S^2 and pasting in their place k Möbius bands. Since a Möbius band is $\mathbb{R}P^2$ minus a disk we deduce that N_k is k $\#$ $\mathbb{R}P^2$, i.e. the connected sum of k copies of $\mathbb{R}P^2$ (see [R]).

Note that

$$\mathbb{R}P^2 \# \mathbb{R}P^2 = K^2$$

is the Klein bottle (Fig. 20). Also

$$3 \# \mathbb{R}P^2 = F_1 \# \mathbb{R}P^2 \ ,$$

because

$$K^2 \# B \cong F_1 \# B \ ,$$

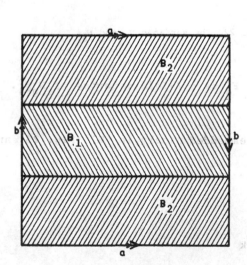

Fig. 20. The Klein bottle divided into two Möbius bands

where B is the Möbius band (Fig. 21). Hence N_k is

$$F_g \ \# \ \mathbb{R}P^2 \quad (k = 2g + 1)$$

or

$$F_g \ \# \ K^2 \quad (k = 2g + 2)$$

(Fig. 22).

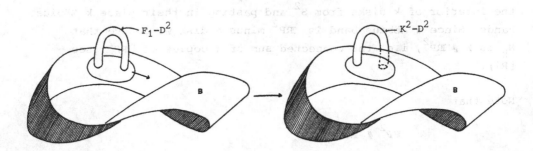

Fig. 21. The torus becomes a Klein bottle

Fig. 22. The surfaces N_7 and N_8.

Exercise 1. Show that F_g is a 2-fold covering of N_{g+1}.

<u>Hint</u>. Think of F_g as embedded nicely in $\mathbb{R}^3$ (Fig. 23) and reflect through the origin.

1.7 S^1-bundles over N_k

We now extend the results of 1.4 to orientable S^1-bundles over non-orientable surfaces. We will show that they are also charac-

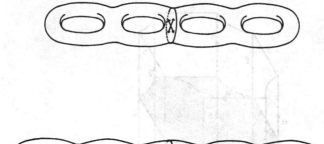

Fig. 23. Involutions in F_4 and F_5 that reverse orientation

terized up to orientation-preserving equivalence by the Euler number.

We have seen in the last section that we can cut open N_k along one curve to get an orientable surface with boundary (Fig. 22). Doing this and then cutting open the resulting orientable surface, we obtain a polygon, in which we have to identify pairs of sides to obtain N_k. All pairs of sides, except one, are pasted exactly as in the orientable case. However, precisely two edges b and b' are pasted in the "other way" (Fig. 20). Thus the *orientable* S^1-bundle is perfectly defined outside of a disk $U \subset N_k$ (the identification of $S^1 \times b$ and $S^1 \times b'$ is done as in Fig. 24).

The definition of the Euler number e is now exactly the same as in the base-orientable case. The Euler number measures the obstruction to the existence of a section (e = 0 if and only if the S^1-bundle admits a section).

We denote by (O n k | -e) the orientable (O) S^1-bundle with non-orientable base (n) of genus k and Euler class e.

Exercise 1. Three different sections of (O n k | 0) *intersect each other.*

Exercise 2. Construct the "quaternionic space" (On1|2).

28

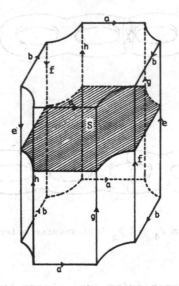

Fig. 24. S^1-bundle over K^2-disk (S = section)

Exercise 3. *Find presentations for the fundamental groups of*
(Oog|b) *and* (Onk|b).

Exercise 4. *Prove that* (Oog|2b) *is a two-fold covering of*
(On g+1|b).

<u>Solution</u>.- Delete from F_g two disks D_1, D_2 corresponding under the
involution u of Fig. 23, obtaining P. Paste to $S^1 \times P$ two solid
tori with meridians $\partial D_1 + bH_1$ and $\partial D_2 + bH_2$ which correspond to
each other via (-1) × u. The argument used to prove the main result
of 1.4 shows that the resulting manifold is (Oog|2b). It is enough
now to extend the involution in the obvious way.

The Euler number of (Onk|e) also has an interpretation similar
to that given in 1.4. Thus, *if ξ is a D^2-bundle over N_k with
orientable total space W^4, the Euler number of the S^1-bundle ∂W^4 is
$N_k \cdot N_k$ in W^4.*

The proof of this is clear once some remarks are made. Note that
$N_k \cdot N_k$ makes sense. In fact, move N_k a little to obtain N_k' in general
position with respect to N_k. After the move, any local orientation of
N_k defines one in N_k'. In a point

$$x \in N_k \cap N_k'$$

we take vectors v_1, v_2 (in N_k) and v_1', v_2' (in N_k') in such a way that the local orientations (x, v_1, v_2), (x, v_1', v_2') are preserved by the perturbation. The number $\varepsilon(x)$ assigned to x is +1 or -1 according to whether or not the orientation $(x, v_1, v_2, v_1', v_2')$ coincides with the orientation of W^4. The number $\varepsilon(x)$ does not depend on the order or on the local orientations selected. We define $N_k \cdot N_k$ as

$$\Sigma \, \varepsilon(x), \quad x \in N_k \cap N_k' \quad .$$

Now an adaptation of the analogous claim in 1.4 completes the proof.

Since the spherical tangent bundle $ST(N_k)$ coincides with the spherical normal bundle $SN(\Delta)$ of the diagonal Δ of $N_k \times N_k$, and the argument of 1.4 shows that $\Delta \cdot \Delta$ is $\chi(N_k)$, we have that *the spherical tangent bundle* $ST(N_k)$ *over a non-orientable surface* N_k *is* $(0 \, n \, k \mid -\chi(N_k))$, i.e. *its Euler number coincides with the Euler characteristic* 2-k *of* N_k.

Exercise 5. Prove that $N_k \times N_k$ *is non-orientable but that a tubular neighbourhood of* $N(\Delta)$ *in* $N_k \times N_k$ *is orientable.*

Hint.- $N_k \times D^2 \subset N_k \times N_k$ is non-orientable, but Δ can be covered with local charts $U \times U$.

Exercise 6. Use exercise 4 to give a different proof of the last result.

Hint.- The 2-fold covering

$$f: F_g \to N_{g+1}$$

induces the 2-fold covering

$$Tf: T(F_g) \to T(N_{g+1}) \quad .$$

Hence

$$ST(N_{g+1}) = (On\ g+1|-\chi(F_g)/2) = (On\ g+1|g-1)\ .$$

1.8 An illustrative example: $\mathbb{RP}^2 \subset \mathbb{CP}^2$

$\mathbb{RP}^2$ is a non-orientable surface embedded in $\mathbb{CP}^2$. The latter is oriented as a complex manifold: in the complex vector space $T_x(\mathbb{CP}^2)$ take a basis (v_1, v_2) and orient $T_x(\mathbb{CP}^2)$ by means of the ordered vectors (v_1, iv_1, v_2, iv_2).

Taking homogeneous coordinates $(z_1 : z_2 : z_3)$ for $\mathbb{CP}^2$, $z_i \in \mathbb{C}$, we consider the *absolute* A^2 of $\mathbb{CP}^2$, i.e. the algebraic curve of degree 2 (conic) given by the equation

$$z_1^2 + z_2^2 + z_3^2 = 0\ .$$

Note that $A^2 \cap \mathbb{RP}^2$ is the empty set.

A^2 is an S^2. This is clear because A^2 is a non-degenerate conic. But here is the proof. Let $(x_1 : x_2 : x_3)$ and $(y_1 : y_2 : y_3)$ be different points of $\mathbb{RP}^2$. Then the set

$$\{(x_1 + \lambda y_1 : x_2 + \lambda y_2 : x_3 + \lambda y_3),\ \lambda \in \mathbb{R} + \infty\}$$

is a $\mathbb{RP}^1$ in $\mathbb{RP}^2$ (indeed, it is the real projective line containing $(x_1 : x_2 : x_3)$ and $(y_1 : y_2 : y_3)$). This $\mathbb{RP}^1$ is contained in the $\mathbb{CP}^1$ given by

$$\{(x_1 + \lambda y_1 : x_2 + \lambda y_2 : x_3 + \lambda y_3),\ \lambda \in \mathbb{C} + \infty\},$$

which intersects A^2 in two points by Bezout's theorem (see [BK] or [Fu]). In fact, the points are distinct, since

$$(x_1 + \lambda y_1)^2 + (x_2 + \lambda y_2)^2 + (x_3 + \lambda y_3)^2 = 0$$

can be rewritten as

$$x_1^2 + x_2^2 + x_3^2 + 2\lambda(x_1 y_1 + x_2 y_2 + x_3 y_3) + \lambda^2(y_1^2 + y_2^2 + y_3^2) = 0$$

whose discriminant is

$$(x_1 y_1 + x_2 y_2 + x_3 y_3)^2 - (x_1^2 + x_2^2 + x_3^2)(y_1^2 + y_2^2 + y_3^2) < 0$$

by the Schwarz inequality, because (x_1, x_2, x_3) and (y_1, y_2, y_3) are independent in $\mathbb{R}^3$. Thus $\mathbb{C}P^1 \cap A^2$ are two points that correspond under the involution

$$J : \mathbb{C}P^2 \to \mathbb{C}P^2$$

given by

$$(z_1 : z_2 : z_3) \mapsto (\bar{z}_1 : \bar{z}_2 : \bar{z}_3) .$$

Hence A^2 is a 2-fold covering of the manifold of projective lines in $\mathbb{R}P^2$.

By duality in $\mathbb{R}P^2$, that manifold is again $\mathbb{R}P^2$ so we can conclude that $A^2 \cong S^2$ (Fig. 25).

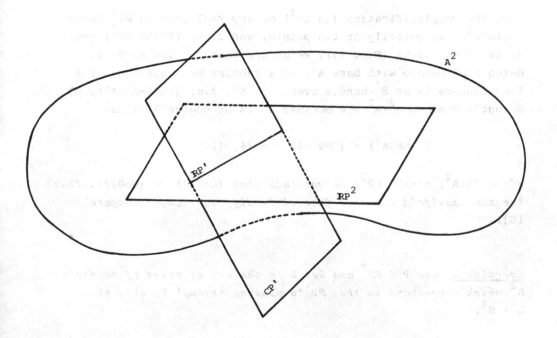

Fig. 25. The absolute of $\mathbb{C}P^2$

Let $N(\mathbb{RP}^2)$ be a tubular neighbourhood of $\mathbb{RP}^2$ in $\mathbb{CP}^2$. The boundary of $N(\mathbb{RP}^2)$ is the spherical normal bundle of $\mathbb{RP}^2$, i.e. $(0\ n\ 1\ |\ -\ \mathbb{RP}^2 \cdot \mathbb{RP}^2)$. To obtain $\mathbb{RP}^2 \cdot \mathbb{RP}^2$ take a tangent vector field $\vec{v}$ on $\mathbb{RP}^2$. Since

$$\chi(\mathbb{RP}^2) = 1$$

we may assume $\vec{v}$ has just one isolated zero of index +1. We can push $\mathbb{RP}^2$ out along the normal vector field $i\vec{v}$ to obtain $\mathbb{RP}_1^2$. Then

$$\mathbb{RP}_1^2 \cdot \mathbb{RP}^2 = -1$$

(the intersection is the zero of $\vec{v}$). Hence $SN(\mathbb{RP}^2)$ is $(0\ n\ 1\ |\ 1)$, which equals $ST(\mathbb{RP}^2)$, as we have seen in 1.7.

Exercise 1. Check that $\mathbb{RP}_1^2 \cdot \mathbb{RP}^2 = -1$.

Now, the complexification (in $\mathbb{CP}^2$) of any real line in $\mathbb{RP}^2$ intersects A^2 transversally in two points, and since $\mathbb{CP}^1 - N(\mathbb{RP}^2)$ consists of two disks (Fig. 26), we deduce that $\mathbb{CP}^2 - \text{Int } N(\mathbb{RP}^2)$, being a D^2-bundle with base A^2, is a tubular neighbourhood of A^2. Its boundary is an S^1-bundle over $A^2 \cong S^2$, i.e. $(0o0\,|\,-A^2 \cdot A^2)$. By Bezout's Theorem $A^2 \cdot A^2 = 4$ because A^2 is of degree 2. Thus

$$SN(A^2) = (0o0\,|\,-4) = -L(4, 1).$$

Since $SN(A^2) = -SN(\mathbb{RP}^2)$ *we conclude that* $(0n1\,|\,1) \cong (0o0\,|\,4)$. *Thus the same manifold might be fibered in different ways* (compare [S]).

Exercise 2. Let $P \in \mathbb{RP}^2$ *and let* L *be the set of pairs of points of* A^2 *which correspond to the* $\mathbb{RP}^1$'s *passing through* P. *Show that* $L \cong S^1$.

Solution. Take vectors v, tangent to $\mathbb{RP}^2$ at P. The vectors iv "point" to the points of L. Identifying P with the origin of

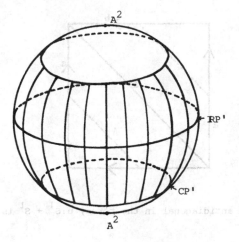

Fig. 26. Intersection of $\mathbb{C}P^2$ with $N(\mathbb{R}P^2)$

coordinates of $\mathbb{C}^2$ and part of $\mathbb{R}P^2$ with $\mathbb{R}^2 \subset \mathbb{C}^2$, we deduce that
the iv's parametrize a circle in the purely imaginary part of $\mathbb{C}^2$.
(This exercise illustrates projective duality, and is due to
Alexis Marin.)

1.9 The projective tangent S^1-bundles

Let u be a fixed point free involution of F_g with quotient space
N_{g+1} (as in Exercise 1, p. 34). We want to make a general construction similar to the one of the last section.

Take $F_g \times F_g$, the diagonal Δ and the antidiagonal

$$\Lambda = \{ (x, ux) \in F_g \times F_g \} \ .$$

Since u is free (i.e. has no fixed points), $\Delta \cap \Lambda$ is empty (Fig. 27).

The involution of $F_g \times F_g$ defined by $(x, y) \mapsto (y, u(x))$ inverts the
orientation and sends Δ onto Λ, showing that

$$\Lambda \cdot \Lambda = -\Delta \cdot \Delta = -\chi(F_g) = 2g-2 \ .$$

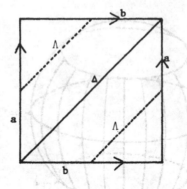

Fig. 27. Diagonal and antidiagonal in the torus; $u:S^1 \to S^1$ is 180° rotation

Let

$$I:F_g \times F_g \to F_g \times F_g$$

be the involution defined by

$$I(x, y) = (y, x) .$$

Here the set of fixed points is Δ, and if C^4 is the quotient space of $F_g \times F_g$ under I, we have a 2-fold covering

$$i:F_g \times F_g \to C^4$$

branched over

$$i(\Delta) \cong \Delta .$$

The image of Λ is

$$F_g/u \cong N_{g+1} \qquad \text{(Fig. 28)}.$$

Exercise 1. Prove the last claim.

The involution I restricted to $SN(\Delta)$ rotates each disk transverse to Δ by 180°, and since

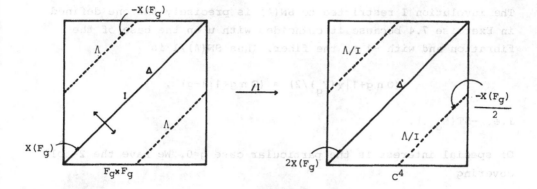

Fig. 28. The involution I

$$SN(\Delta) \cong ST(F_g),$$

we deduce that $SN(\Delta)/I$, which coincides with $SN(i(\Delta))$, is $PT(F_g)$. Hence we have

$$PT(F_g) = (Oog\,|-2\chi(F_g)) = (Oog\,|4g-4),$$

to see this it is enough to understand the behaviour of I with respect to the circle

$$C \sim m + \chi(F_g)\ell$$

introduced in section 1.2 (Fig. 29).

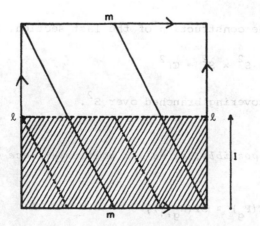

Fig. 29. The Euler number doubles, when passing to the quotient (g = 2)

The involution I restricted to SN(Λ) is precisely the one defined in Exercise 7.4 because it coincides with u on the base of the fibration and with -1 in the fiber. Thus SN(Λ)/I is

$$(0 \ n \ g+1 \ | \ \chi(F_g)/2) = (0 \ n \ g+1 \ | \ 1-g) \ ,$$

i.e. $-ST(N_{g+1})$.

Of special interest is the particular case g=0. We have the 2-fold covering

$$i: \ S^2 \times S^2 \rightarrow C^4 \ ,$$

branched over

$$i(\Delta) \cong S^2 \ .$$

The manifold Λ/I is $\mathbb{R}P^2$ and

$$C^4 = N(S^2) \cup_\partial N(\mathbb{R}P^2) \ ,$$

hence

$$PT(S^2) = ST(\mathbb{R}P^2) = L(4,1) \ .$$

Since in Exercise 5 we will see that

$$C^4 \cong \mathbb{C}P^2 \ ,$$

the case g=0 is the construction of the last section. Note that

$$i: \ S^2 \times S^2 \rightarrow \mathbb{C}P^2$$

is then a 2-fold covering branched over S^2.

Exercise 2. Is it possible, as in the particular case g=0, to conclude that

$$PT(F_g) \cong ST(N_{g+1}),$$

for g > 0?

<u>Solution</u>.- No,(compare $H_1(PT(F_1); \mathbf{Z})$ with $H_1(ST(N_2); \mathbf{Z})$). In this case C^4 is not

$$N(i(\Delta)) \cup_{\partial} N(i(\Lambda))$$

(why?).

Besides I, we have the involutions J and K in $F_g \times F_g$ defined by

$$J(x, y) = (ux, uy)$$

and

$$K(x, y) = (uy, ux) \quad .$$

Thus the Klein group $\mathbf{Z}_2 \times \mathbf{Z}_2$ acts in $F_g \times F_g$. We have the following conmutative diagram among coverings

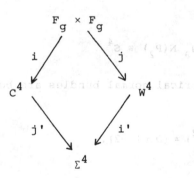

Here all coverings are branched except j. Here $j(\Lambda)$ and $j(\Delta)$ have the same spherical normal bundles, up to orientation, namely

$$(O \, n \, g+1 \, | -\chi(F_g)/2),$$

i.e. $ST(N_{g+1})$ (compare this with 1.7). The 2-fold covering

$$j':C^4 \to \Sigma^4$$

is branched over $j'(N_{g+1})$, whose spherical normal bundle is, up to orientation

$$(O \, n \, g+1 \mid 2 \, \frac{-\chi(F_g)}{2}) = (O \, n \, g+1 \mid 2g-2)$$

(Fig. 29). Hence we have

$$PT(N_{g+1}) = (O \, n \, g+1 \mid 2g-2) \ .$$

The spherical normal bundle $SN(j'i(\Delta))$ is also

$$(O \, n \, g+1 \mid -\frac{1}{2}(2\chi(F_g))) = (O \, n \, g+1 \mid -\chi(F_g)) \ .$$

When g=0, we will see in Exercise 5 that

$$\Sigma^4 \cong S^4 \ .$$

In S^4 there are two (disjoint) real projective planes, P_1 and P_2, such that

$$N(P_1) \ \cup_\partial \ N(P_2) \cong S^4$$

and the associated spherical normal bundles are both equal to (the quaternionic space)

$$PT(\, \mathbb{R}P^2) = (O \, n \, 1 \mid -2) \ .$$

Exercise 3. *Prove that W^4 is an (orientable) bundle with base and fiber N_{g+1}. Study $i':W^4 \to \Sigma^4$.*

Exercise 4. *Prove that $(F_g \times F_g)/K \cong -C^4$.*

Solution.- $\varphi:(x, y) \longmapsto (x, uy)$ inverts orientation in $F_g \times F_g$ and is such that $\varphi I = K\varphi$.

Exercise 5. *Determine C^4, W^4, $(S^2 \times S^2)/K$ and Σ^4, for g=0.*

Solution.- C^4 is $\mathbb{C}P^2$. To see this, take homogeneous coordinates in $\mathbb{C}P^3$, $(z_1 : z_2 : z_3 : z_4)$, and identify $\mathbb{C}P^2$ with the algebraic variety $z_4 = 0$. The absolute A^4 of $\mathbb{C}P^3$ is defined by the equation

$$z_1^2 + z_2^2 + z_3^2 + z_4^2 = 0 .$$

Then

$$A^4 \cap \mathbb{C}P^2 = A^2 \cong S^2$$

(compare section 1.8 and see Fig. 30).

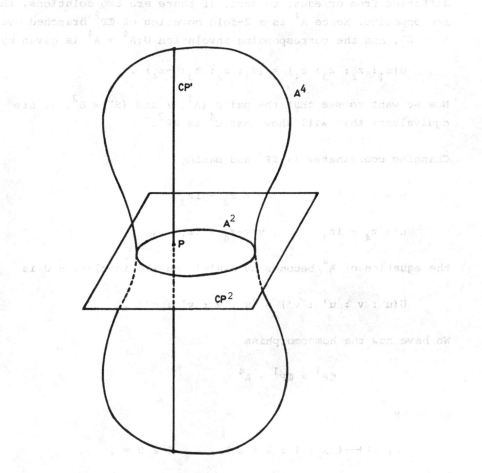

Fig. 30. The absolute of $\mathbb{C}P^3$

Each point $P = (a_1 : a_2 : a_3 : 0)$ of $\mathbb{C}P^2$ determines the line

$$\mathbb{C}P^1 = \{(a_1 : a_2 : a_3 : \lambda); \ \lambda \in \mathbb{C}\} .$$

This line intersects A^4 in two points or one point according to whether or not P is on $A^2 = S^2$, because the equation

$$a_1^2 + a_2^2 + a_3^2 + \lambda^2 = 0$$

has two solutions, or one, according to whether $a_1^2 + a_2^2 + a_3^2$ is different from or equal to zero. If there are two solutions, they are opposite. Hence A^4 is a 2-fold covering of $\mathbb{C}P^2$ branched over $A^2 \cong S^2$, and the corresponding involution $U:A^4 \to A^4$ is given by

$$U(z_1 : z_2 : z_3 : z_4) = (z_1 : z_2 : z_3 : -z_4) \ .$$

Now we want to see that the pairs (A^4, U) and $(S^2 \times S^2, I)$ are equivalent; this will show that C^4 is $\mathbb{C}P^2$.

Changing coordinates in $\mathbb{C}P^3$ and making

$$u = z_2 + iz_3 \qquad \qquad v = z_2 - iz_3$$

$$u' = z_4 + iz_1 \qquad \qquad v' = -z_4 + iz_1 \quad ,$$

the equation of A^4 becomes $uv = u'v'$ and the involution U is

$$U(u : v : u' : v') = (u : v : v' : u') \ .$$

We have now the homeomorphism

$$\mathbb{C}P^1 \times \mathbb{C}P^1 \to A^4$$

given by

$$(\lambda, \mu) \longmapsto (\lambda\mu : 1 : \lambda : \mu) \ , \quad \lambda, \mu \in \mathbb{C} \cup \infty \ ,$$

with inverse given by

$$(u : v : u' : v') \longmapsto (\tfrac{u}{u'} , \tfrac{u}{v'}) \ .$$

The involution U is then

$$(\lambda, \mu) = (\tfrac{u}{u'} , \tfrac{u}{v'}) \longmapsto (\tfrac{u}{v'} , \tfrac{u}{u'}) = (\mu, \lambda)$$

which is I. This proves $C^4 \cong \mathbb{C}P^2$.

The pair (A^4, V), where V is given by

$$(z_1 : z_2 : z_3 : z_4) \longmapsto (\bar{z}_1 : \bar{z}_2 : \bar{z}_3 : \bar{z}_4) \ ,$$

is $(S^2 \times S^2, J)$. In fact, in coordinates $(u : v : u' : v')$ we have

$$V(u : v : u' : v') = (\bar{v} : \bar{u} : -\bar{v}' : -\bar{u}') \ .$$

Hence V sends

$$(\lambda, \mu) = (\frac{u}{u'}, \frac{u}{v'})$$

into

$$(\frac{\bar{v}}{-\bar{v}'}, \frac{\bar{v}}{-\bar{u}'}) = (-\frac{1}{\lambda}, -\frac{1}{\mu}) \ ,$$

and $\lambda \longmapsto -\frac{1}{\lambda}$ is equivalent to the antipodal map in $\mathbb{C}P^1 \cong S^2$. The manifold

$$W^4 = (S^2 \times S^2)/J$$

is A^4/V which is homeomorphic to the space of projective lines of CP^3 (= the space of planes in the vector space $\mathbb{C}^4$) (compare 1.8). Finally (A^4, Y), where Y sends $(z_1 : z_2 : z_3 : z_4)$ to $(\bar{z}_1 : \bar{z}_2 : \bar{z}_3 : z_4)$, is equivalent to $(S^2 \times S^2, K)$. The antidiagonal $\Lambda \subset A^4$ is then

$$\{(x_1 : x_2 : x_3 : \lambda) \,|\, (x_1, x_2, x_3) \in \mathbb{R}^3, \ \lambda \in \mathbb{C}, \ y - \lambda^2 = x_1^2 + x_2^2 + x_3^2\} \ .$$

We already know (Exercise 4) that $(S^2 \times S^2)/K = -\mathbb{C}P^2$.

It only remains to see that Σ^4 is S^4, i.e. that $\mathbb{C}P^2/J$, where

$$J(z_1 : z_2 : z_3) = (\bar{z}_1 : \bar{z}_2 : \bar{z}_3) \ ,$$

is S^4. From 1.5 we deduce that $\mathbb{C}P^2$ is $D^4/\sim$, where $\sim$ collapses fibers of the Hopf fibration of ∂D^4. The homeomorphism J is induced by 180° rotation of D^4 around a disk, whose boundary is the circle $Oe_2 \subset S^3$ (Fig. 16). The quotient of D^4 under such a rotation is again D^4 and ∂D^4 appears "fibered" as indicated in Figures 31 and 32 by the images of the fibers of the Hopf fibration. It is necessary to prove that D^4/R, where R collapses these new "fibers"

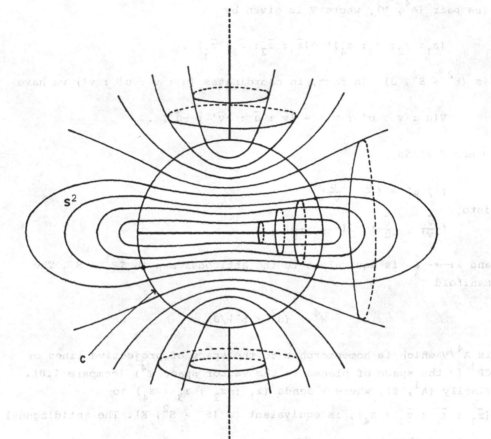

Fig. 31. S^3 decomposed into spheres, the images under J of the tori in Figure 18, two of which are degenerate. The circle C is the image of Oe_2

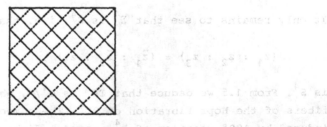

Fig. 32. "Fibration" of the 2-sphere S^2 of Fig. 31 by circles. There are two degenerate fibers

to points, is S^4. After collapsing the degenerate fibers, we obtain the "fibration" of Figure 33. It is now clear that by collapsing the fibers of Fig. 33 we obtain S^4.

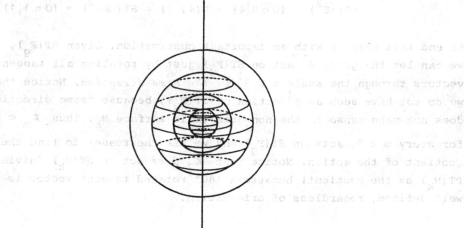

Fig. 33. The fibers are the circles and the points of the axis

Exercise 6. *Find the spherical normal bundle of an* $\mathbb{R}P^2$ *standardly embedded in* S^4 *(i.e. cap off a Möbius band in* $\mathbb{R}^3$ *by a (smooth) disk in* $\mathbb{R}^4$ *, smoothly glued to* ∂*(Möbius band)).*

Answer. $(0 \, n \, 1 | \pm 2)$.

Exercise 7. *Show that the degenerate fibers of the "fibration" of* ∂D^4 *shown in Figures 31, 32 constitute a Möbius band, whose boundary is the circle C of Figure 31.*

Summing up, we have seen that

$$ST(F_g) = (0 \circ g | -2 + 2g) \; ;$$

$$PT(F_g) = (O \circ g \mid -4 + 4g) \; ;$$

$$ST(N_k) = (O \, n \, k \mid -2 + k) \; ;$$

$$PT(N_k) = (O \, n \, k \mid -4 + 2k) \; .$$

Moreover

$$-PT(S^2) = (O \circ 0 \mid 4) = L(4, 1) = ST(\, \mathbb{R}P^2) = (O \, n \, 1 \mid 1) .$$

We end this chapter with an important observation. Given $ST(F_g)$, we can let the group S^1 act on $ST(F_g)$ just by rotating all tangent vectors through the angle $\sigma \in S^1$ *in the same direction*. Notice that we do *not* have such an S^1-action on $ST(N_k)$, because "same direction" does not make sense in the non-orientable surface N_k. Thus $\mathbb{Z}_m \subset S^1$, for every $m \geq 1$, acts on $ST(F_g)$ and we ask the reader to find the quotient of the action. Notice that $\mathbb{Z}_2$ *does* act in $ST(N_k)$ (giving $PT(N_k)$ as the quotient) because a $180°$ rotated tangent vector is well defined, regardless of orientation.

Exercise 8. *Show that there is a 3-fold covering* $(O \, n \, 1 \mid 3) \rightarrow (O \, n \, 1 \mid 1)$ *which is fiber preserving (the covering must be irregular).*

Exercise 9. *Remember that the Klein bottle K is a S^1-bundle over S^1. Show that there is a 3-fold covering $K \rightarrow K$ which is fiber preserving. Is it regular? Find its monodromy (see [ST] for the definition of monodromy and for coverings).*

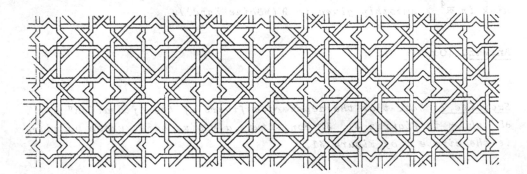

Chapter Two: Manifolds of Tessellations on the Euclidean Plane

> "Digo, pues, que encima del patio de nuestra
> prisión caían las ventanas de la casa de un Moro
> rico, y principal; las cuales como de ordinario
> son las de los Moros, más eran agujeros que
> ventanas, y aun éstas se cubrían de zelosías muy
> espesas y apretadas."
>
> *"Now, overlooking the courtyard of our prison were
> the windows of the house of a rich and important
> Moor, which, as is usual in Moorish houses, were
> more like loopholes than windows, and even so were
> covered by thick and close lattices."*
> *Cervantes, Don Quixote, Part I, Ch. XL, The
> Captive's Story continued.*

The space whose points are the different positions of a regular
dodecahedron inscribed in the 2-sphere, with the most natural
possible topology, is a closed, orientable 3-manifold known as the
Poincaré homology 3-sphere. A dodecahedron is a tessellation of the
2-sphere, as is an octahedron or a tetrahedron, for instance. The
original examples of tessellations belong to the euclidean plane
$\mathbb{R}^2$, like the hexagonal mosaics that one can admire in *The Alhambra
de Granada* or in *The Aljafería de Zaragoza*. The hyperbolic plane
H^2 is very rich in tessellations. The object of the rest of the
book is to describe the 3-manifolds of euclidean, spherical and
hyperbolic tessellations.

This chapter, the next, and Chapter Five are natural continuations
of Chapter One. In Chapter One we studied spherical tangent bundles
of closed surfaces. Here we are studying tangent bundles of orbifolds. In fact any tessellation t in $X = \mathbb{R}^2$, S^2 or H^2 defines a
group Γ of orientation-preserving symmetries fixing it as a set. The
quotient X/Γ is a 2-dimensional orbifold. In this context, the manifold of the tessellation t is the spherical tangent bundle of the
orbifold X/Γ. In Chapters Two, Three and Five we study the euclidean,
spherical and hyperbolic cases of this situation.

The rigorous definition of a tessellation will be introduced in
section 2.6 after we have studied a concrete tessellation in
sections 2.1 to 2.5. The latter tessellation is a tiling in the
euclidean plane E^2 made of squares of side one and we request the
reader's indulgence in having to wait till section 2.6 for the
definition.

Appendix A is devoted to the definition of an orbifold and of the
spherical tangent bundle of an orbifold. Readers who are not
acquainted with these concepts should read sections A.1, A.2 and
A.3 before starting this chapter.

2.1 The manifold of square-tilings

A *square-tiling* (or simply a *tiling*) will here be a tiling of the
euclidean plane E^2 by square tiles of side one. More rigorously,
a square-tiling is a subset of E^2 which is the union of two sets
of equally spaced parallel lines, such that lines belonging to
different sets are perpendicular (Figure 1).

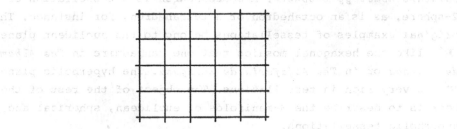

Fig. 1. A square-tiling

Let M be the set of square-tilings of the euclidean plane E^2. We
can parametrize M by points of $\mathbb{R}^2 \times \mathbb{R}$ as follows. The point
$(v, \alpha) \in \mathbb{R}^2 \times \mathbb{R}$ determines the square-tiling with a corner in v
and such that if it is rotated in the angle $-\alpha$, its families of
parallel lines become parallel to the axes of E^2 (Fig. 2). We

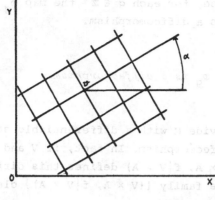

Fig. 2. The square-tiling f(v, α)

give to M the quotient topology of the map $f: \mathbb{R}^2 \times \mathbb{R} \to M$ just defined, i.e. a set $U \subset M$ is open if and only if $f^{-1}(U)$ is open in $\mathbb{R}^2 \times \mathbb{R}$.

The set of elements of $\mathbb{R}^2 \times \mathbb{R}$ which project onto f(v, α) is

$$\{(v + m(\cos \alpha', \sin \alpha') + n(-\sin \alpha', \cos \alpha'), \alpha + p \frac{\pi}{2}) :$$

$$(m, n, p) \in \mathbb{Z}^3, 0 \leq \alpha' < \frac{\pi}{2}, \alpha' \equiv \alpha \bmod \frac{\pi}{2}\} .$$

Thus M is just the set of orbits of the action φ of $\mathbb{Z}^3$ in $\mathbb{R}^3$ given by

$$[(m, n, p), (v, \alpha)] \longmapsto (v + m(\cos \alpha', \sin \alpha')$$

$$+ n(-\sin \alpha', \cos \alpha'), \alpha + p \frac{\pi}{2}) ,$$

with the quotient topology. This action is *free* (the identity e is the unique element of $\mathbb{Z}^3$ fixing a point) and *properly-discon-tinuous*, because every $(v, \alpha) \in \mathbb{R}^3$ has a neighbourhood $U \subset \mathbb{R}^3$ such that

$$U \cap \varphi(g, U) = \emptyset$$

for every $g \neq e$, $g \in \mathbb{Z}^3$ (take U to be $V \times A$, where V is an open disk of radius $< \frac{1}{2}$ and centered in v, and A is an open interval $< \frac{\pi}{2}$, centered in α). Moreover, this action of $\mathbb{Z}^3$ on $\mathbb{R}^3$ is *differentiable*, because for each $g \in \mathbb{Z}^3$ the map $\varphi_g : \mathbb{R}^3 \to \mathbb{R}^3$ given by $r \longmapsto \varphi(g,r)$ is a diffeomorphism.

Exercise 1. *Show that* φ_g *is a diffeomorphism*.

Now it is easy to provide M with a differentiable structure so that f is a local diffeomorphism. In fact, if V and A are as before, the set of charts $(V \times A, f|V \times A)$ defines this differentiable structure, because the family $\{(V \times A, f|V \times A)\}$ clearly covers M, and if

$$q \in (f|V_1 \times A_1)(V_1 \times A_1) \cap (f|V_2 \times A_2)(V_2 \times A_2) \neq \emptyset$$

and $s \in V_1 \times A_1$, $r \in V_2 \times A_2$ are such that

$$f(s) = f(r) = q \quad ,$$

there is a $g \in \mathbb{Z}^3$ with $\varphi_g(r) = s$. Thus, the map

$$(f|V_1 \times A_1)^{-1} \cdot (f|V_2 \times A_2)$$

coincides with φ_g in a neighbourhood of r, and it is differentiable. The manifold M is thus a differentiable manifold without boundary, f is a local diffeomorphism, and, in fact, f is the universal covering of M. Moreover M is compact, because an infinite sequence of tilings must have an infinite number of vertices in a tile, which would in turn have an accumulation point P contained in the tile; therefore, the tilings of the sequence accumulate in a tiling of vertex P, because the directions of their lines vary on S^1, which is compact.

In a similar way, given a particular pattern such as the one of Fig. 3, we can create the *3-manifold of positions* of that pattern in E^2. How many closed, orientable 3-manifolds can we obtain using this procedure? It turns out that the 3-manifolds corresponding to the patterns of Figures 1 and 3 are the same! This is because, as intuition immediately suggests, the particular design of the pattern is not relevant; what is important is its *rigid symmetry*, and it is an interesting exercise to check that the beautiful

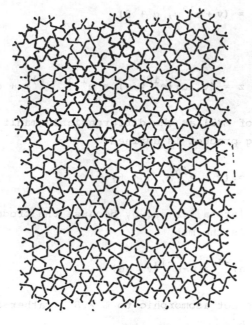

Fig. 3. A symmetric pattern. (From G. Schneider "Geometrische Bauornamente der
Seldschuken in Kleinasien" L. Reichert Verlag Wiesbaden (1980))

pattern of Fig. 3 has the same rigid symmetry features as the
square-tiling of Fig. 1 (though the latter also admits *reflection
symmetries*, which are not present in the former). To see all this
we need a little interlude, to talk about the group of isometries
of E^2.

2.2 The isometries of the euclidean plane

The group $E(E^2) := E(2)$ of isometries of the euclidean plane which
preserve orientation consists of the transformations

$$(v, \alpha) : \mathbb{C} \longrightarrow \mathbb{C}$$

$$z \longmapsto (v, \alpha)z := e^{i\alpha}z + v$$

where $v \in \mathbb{C}$ and $\alpha \in \mathbb{R}/2\pi\mathbb{Z} = S^1$. The product in $E(2)$ is composition
of transformations. Hence we have

$$(v, \alpha)(w, \beta) = (v + e^{i\alpha}w, \alpha + \beta)$$

because

$$(v, \alpha)(w, \beta) z = (v, \alpha)(e^{i\beta}z + w) = e^{i(\alpha + \beta)}z + e^{i\alpha}w + v \ .$$

Thus we can think of E(2) as the differentiable manifold $\mathbb{C} \times S^1$ in which the following product is defined:

$$(1) \qquad (v, \alpha)(w, \beta) = (v + e^{i\alpha}w, \alpha + \beta)$$

This converts E(2) into a *Lie group*, because the product (1) is differentiable, as is the map

$$(v, \alpha) \longmapsto (v, \alpha)^{-1} = (-e^{-i\alpha}v, -\alpha)$$

As a group, E(2) is not isomorphic to $\mathbb{C} \times S^1$, rather it is a semi-direct product $\mathbb{C} \ltimes S^1$.

Exercise 1. Prove the last assertion. Find the action of S^1 on $\mathbb{C}$.

Hint.- S^1 acts on $\mathbb{C}$, conjugating the translations: $(v \longmapsto e^{i\alpha} v)$.

The universal cover of E(2) is the Lie group $\tilde{E}(2) = \mathbb{C} \times \mathbb{R}$ with the product defined by

$$(v, r)(w, s) = (v + e^{ir}w, r + s).$$

Exercise 2. Prove that the commutator subgroup of E(2) *equals the subgroup of translations.*

Hint.-

$$(v, \alpha)(w, \beta)(v, \alpha)^{-1}(w, \beta)^{-1} = (v + e^{i\alpha}w, \alpha + \beta)(-e^{-i\alpha}v, -\alpha)$$

$$(-e^{-i\beta}w, -\beta) = (v + e^{i\alpha}w, \alpha + \beta)(-e^{-i\alpha}v - e^{-i(\alpha + \beta)}w, -(\alpha+\beta)) =$$

$$= (v + e^{i\alpha}w - e^{i\beta}v - w, 0) = (v(1 - e^{i\beta}) - w(1 - e^{i\alpha}), 0).$$

Define a *pointer* to be any element of the spherical tangent bundle $ST(E^2)$ of E^2 (compare [HC], p. 69); i.e. a pointer will be a vector of length 1 based at some point of E^2 (Fig. 4). Since E^2 is contractible, $ST(E^2)$ is trivial and a trivialization is defined by associating to each pointer a pair $(v, \alpha) \in \mathbb{C} \times S^1$ so that an equivalence between bundles arises. This can be done as follows: $E(2)$ acts on $\mathbb{C}$ and, by the differential, acts on $ST(E^2)$ as well. This action is transitive and the isotropy group of a pointer is trivial. Hence, fixing a pointer b (for instance b = (0,0), Fig. 4) we have the diffeomorphism

$$E(2) \longrightarrow ST(E^2)$$

$$(v, \alpha) \longmapsto (v, \alpha)b$$

and we identify $E(2)$ with $ST(E^2)$ by this diffeomorphism (Fig. 4).

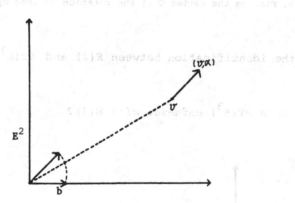

Fig. 4. Pointers b = (0,0) and (v, α)

Exercise 3. Every element $(v, \alpha) \in E(2)$ *can be thought of as a rotation (possibly of infinite radius).*

<u>Hint</u>.- Figure 5.

Exercise 4. The product $(v, \alpha)(w, \beta)$ *is the image of the pointer* (w, β) *under the rotation* (v, α).

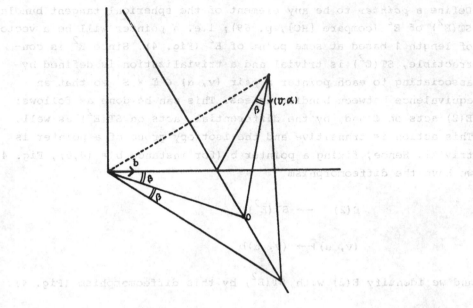

Fig. 5. Finding the center O of the rotation defined by (v, α)

<u>Hint</u>.- Use the identification between E(2) and ST(E^2) (Fig. 6).

<u>*Exercise 6*</u>. *Does* ST(E^3) *coincide with* E(3)?

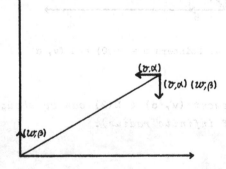

Fig. 6. Product of pointers

2.3 Interpretation of the manifold of square-tilings

While the definition of M, given in 2.1, is very
intuitive (and it is important to always have it in mind), we will
not understand its true significance until we disclose the relationship between M (the set of square-tilings) and the subgroup
$\Gamma \leq E(2)$ of isometries which leave invariant (as a set) a fixed
element C of M.

In fact, E(2) defines a left-action on M:

$$E(2) \times M \longrightarrow M$$

$$((v, \alpha), d) \longmapsto (v, \alpha)d := \{(v, \alpha)z : z \in d\} \quad .$$

It is evident that $(v, \alpha)d \in M$, because (v, α) is an isometry of
E^2 preserving orientation. The action is differentiable, i.e.

$$M \longrightarrow M$$

$$d \longmapsto (v, \alpha)d$$

is a diffeomorphism.

Exercise 1. *Prove the last assertion.*

Since Γ is the isotropy group of the fixed tiling $C \in M$, it follows
that M is diffeomorphic to the homogeneous manifold $E(2)/\Gamma$ of left
cosets $(v, \alpha)\Gamma$ with the quotient differentiable structure, under
the map

$$E(2)/\Gamma \longrightarrow M$$

$$(v, \alpha)\Gamma \longmapsto (v, \alpha)C \quad .$$

Exercise 2. *Prove that $E(2)/\Gamma$ and $\Gamma \backslash E(2) = \{$right cosets$\}$ are*
orientation-reversing diffeomorphic.

<u>Hint.</u>- Project the diffeomorphism $(v, \alpha) = g \mapsto (-ve^{-i\alpha}, -\alpha) = g^{-1}$ of E(2).

It now becomes clear that the manifold made of the patterns given by Fig. 3 is M, because Γ is also the group of isometries which leave the pattern of Fig. 3 invariant as a set.

In the next sections we will study the manifold $\Gamma \backslash E(2)$ which is the same as $E(2)/\Gamma$ but with the opposite orientation. *We orient E(2) in such a way that the identification $E(2) = ST(E^2) = E^2 \times S^1$ is orientation-preserving, where $E^2 \times S^1$ has the product orientation.*

2.4 The subgroup Γ

Γ is clearly generated by the translations $\sigma = (1, 0)$ and $\tau = (i, 0)$ and the rotations $a = (0, \pi/2)$, $b = (1, \pi/2)$ and $c = (1, \pi)$ (Fig. 7).

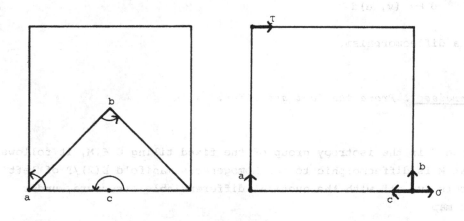

Fig. 7. Generators of Γ

The set of pointers is illustrated in Fig. 8.

Since (Fig. 7) $\sigma a = b$ and $ba = c$, it follows that $\{\sigma, \tau, a\}$ generates Γ.

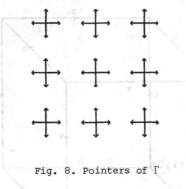

Fig. 8. Pointers of Γ

2.5 The quotient $\Gamma \backslash E(2)$

The manifold $\Gamma \backslash E(2)$ is the result of performing the identifications defined by the left action of Γ in $E(2) = \mathbb{C} \times (\mathbb{R}/2\pi \mathbb{Z})$. We have to identify:

$$(v, \alpha) \equiv \sigma(v, \alpha) = (1, 0)(v, \alpha) = (v + 1, \alpha)$$

$$(v, \alpha) \equiv \tau(v, \alpha) = (i, 0)(v, \alpha) = (v + i, \alpha)$$

$$(v, \alpha) \equiv a(v, \alpha) = (0, \frac{\pi}{2})(v, \alpha) = (e^{i\frac{\pi}{2}}v, \alpha + \frac{\pi}{2})$$

Thus, in $\mathbb{C} \times (\mathbb{R}/2\pi\mathbb{Z})$ we take as a fundamental domain for these identifications a prism with square base of side 1, with vertices $(0, 0)$, $(1, 0)$, $(i, 0)$, $(1 + i, 0)$ and height $\frac{\pi}{2}$ (Fig. 9). The base is pasted to the top square by right rotation of angle $\pi/2$. Each lateral side is identified with its opposite by translation. The manifold M is $\Gamma \backslash E(2)$ with the opposite orientation.

Exercise 1. Construct $E(2)/\Gamma$ in a similar way.

The manifold M is an example of a *mapping torus*: given X and φ, self-homeomorphism of X, the *mapping torus* of X is the quotient $X \times [0, 1] / (x, 0) \sim (\varphi x, 1)$. Therefore the mapping torus of X is the total space of a bundle with fiber X, base S^1 and monodromy φ.

Thus M *is an F_1-bundle over S^1 with periodic monodromy of order 4, where F_1 is the torus.*

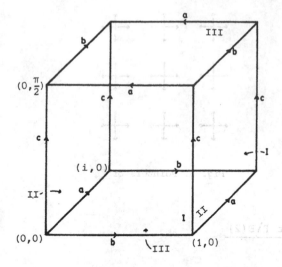

Fig. 9. The manifold M as a prism with identified faces

2.6 The tessellations of the euclidean plane

Given a subset of E^2, the subgroup Γ of E(2) which leaves the subset invariant (as a set) is called *the symmetry group of the subset*. The subset is "asymmetric" when Γ consists only of the identity element. We are interested in the groups of symmetry Γ called *(orientation-preserving) crystallographic groups*; i.e. groups that act on E^2 with discrete orbits, and such that E^2/Γ is compact. This last condition is equivalent to Γ having a fundamental domain of finite area, and it is imposed here in order to ensure that the corresponding manifold of positions of tessellations will be compact. A subset of E^2 whose symmetry group is crystallographic will be called a *tessellation* of E^2. There are many tessellations having the same group (Figures 1 and 3, for instance). Given a subgroup $\Gamma \le E(2)$, the *canonical tessellation* $c(\Gamma)$ will be the set of pointers that define Γ (see Fig. 8). The symmetry group of $c(\Gamma)$ turns out to be Γ, since an element of Γ sends pointers of Γ to pointers of Γ. Conversely if $g \in E(2)$ leaves $c(\Gamma)$ invariant, then $g\gamma \in \Gamma$, for every $\gamma \in \Gamma$, hence $g \in \Gamma$.

In Appendix A we study the crystallographic groups of the euclidean plane and we show that, up to isomorphism, there are only 5 that preserve orientation. These groups are represented in Fig. 10, by

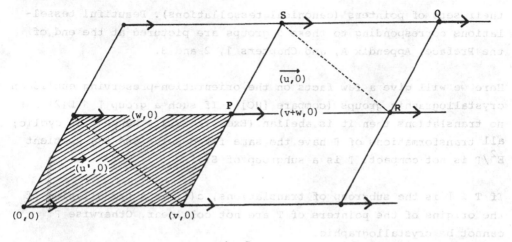

a) S

b) S 2222

c) S 432

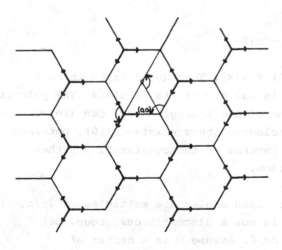

d) S 333

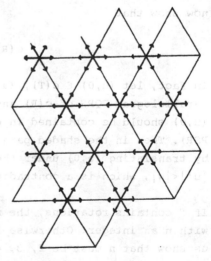

e) S 632

Fig. 10. Orientation-preserving euclidean crystallographic groups

their sets of pointers (canonical tessellations). Beautiful tessellations corresponding to these 5 groups are pictured at the end of
the Preface, Appendix A, and Chapters 1, 2 and 3.

Here we will give a few facts on the orientation-preserving euclidean
crystallographic groups (compare [HC]). If such a group $\Gamma \leq E(2)$ has
no translations then it is abelian (Exercise 2.2), hence Γ is cyclic;
all transformations of Γ have the same fixed-point and the quotient
E^2/Γ is not compact: Γ is a subgroup of S^1.

If $T \leq \Gamma$ is the subgroup of translations, $c(T)$ is a tessellation if
the origins of the pointers of T are not collinear. Otherwise Γ
cannot be crystallographic.

Exercise 1. Prove the last assertion.

Assume then that T contains two pointers which together with the
base pointer (0,0) are not aligned. Let (v,0) be the pointer of T
closest to (0,0). Then $\{(mv,0) : m \in \mathbb{Z}\}$ forms a subgroup of T of
aligned pointers. Among the other pointers of T let (w,0) be the
one closest to (0,0) (Fig. 10a). Then (v,0) and (w,0) generate a
subgroup R of T and let c(R) be the tessellation of Fig. 10a. We
now show that

$$c(R) = c(T) .$$

In fact, let $(u,0) \in c(T)$, $(u,0) \notin c(R)$. Then (u,0) falls in some
parallelogram PQRS of c(R) and is not one of its vertices. The pointer
(u,0) should be contained in one of the triangles PSR, QSR (let be
PSR). Then in the shaded parallelogram there exists (u',0), obtained
by translating (u,0) under the inverse of the pointer P. But then
$|u'|<|w|$, which is a contradiction.

If Γ contains rotations, the rotation angles are multiples of $2\pi/n$,
with n an integer. Otherwise Γ is not a discontinuous group. Let
us show that n must be 2, 3, 4 or 6. Assume O is a center of
$2\pi/n$ rotation with n > 6. Let t be a translation of $T \leq \Gamma$ and let
O' = tO (Fig. 11). Let O" be the result of rotating O' around O
through angle $2\pi/n$, and consider the translations t', t" depicted

in Fig. 11. Since t" is conjugate to t, then t" ∈ Γ. But then
t' ∈ Γ, since t" = t", and |t'|<|t|. Taking t with |t| minimal
leads to a contradiction. That n ≠ 5 is shown in Fig. 12.

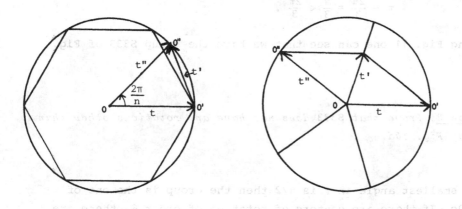

Fig. 11. n must be ≦ 6   Fig. 12. n ≠ 5 because |t'|<|t|

If there is some center of order 2, assume that the base pointer
(0,0) is based at that center P (Fig. 13). The pointer corresponding
to the rotation π with center in P is (0,π). The elements of the
group S2222 are the pointers of Fig. 13. Besides the rotations with
centers in the tessellation c(T) there are three other types of
centers : Q, R, S.

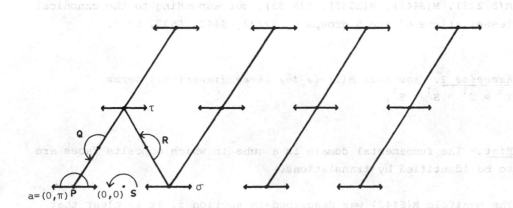

Fig. 13. Generators for S2222

If the smallest angle of rotation in Γ is $2\pi/3$, there cannot exist a center of rotation of order 2. Otherwise, by composition of the rotation of angle π with the one of angle $\frac{-2\pi}{3}$ we would obtain one of angle

$$\pi - \frac{2\pi}{3} = \frac{\pi}{3} < \frac{2\pi}{3} \; .$$

Studying Fig. 11 one can see that we have the group S333 of Fig. 10d.

Exercise 2. Prove that S333 *does not have any rotations other than those of Fig. 10d.*

If the smallest angle in Γ is $\pi/2$ then the group is the one of Fig. 10c. If there are centers of rotation of order 6, there are centers of orders 2 and 3, but not of order 4. The group S632 appears in Fig. 10e. Looking at Fig. 10 it is clear that T ≤ S2222 ≤ S442 and that T ≤ S333 ≤ S632.

2.7 The manifolds of euclidean tessellations

Exactly as we did in section 5 we can find the manifolds M(T), M(S2222), M(S442), M(S632), M(S333), corresponding to the canonical tessellations of the 5 groups T, S2222, S442, S632, S333.

Exercise 1. Show that M(T) *is the three dimensional torus*
$T^3 = S^1 \times S^1 \times S^1$.

<u>Hint</u>.- The fundamental domain is a cube in which opposite faces are to be identified by translations.

The manifold M(S442) was described in section 5. It is clear that we have a four-sheeted covering $T^3 \to M(S442)$ corresponding to T ≤ S442.

Exercise 2. Show that M(S2222) is the result of pasting the bottom and the top faces of a 4-sided prism by a rotation of π and the rest of the faces by translation.

Hint.- Generators for S2222 are σ, τ and a (Fig. 13).

Then we have the 2-sheeted coverings $T^3 \to M(S2222) \to M(S442)$ corresponding to the inclusions $T \leqq S2222 \leqq S442$. *The manifolds M(T), M(S2222), M(S442) are F_1-bundles over S^1 with identity monodromy and periodic monodromy of periods 2 and 4, respectively.*

From Fig. 14 we deduce that $\sigma = (e^{i\frac{\pi}{6}}, 0)$, $\tau = (e^{-i\frac{\pi}{6}}, 0)$, $a = (0, \frac{2\pi}{3})$ generate S333. The action of S333 in $\mathbb{C} \times (\mathbb{R}/2\pi\mathbb{Z})$ is given by the following

$$(v, \alpha) \mapsto (e^{i\frac{\pi}{6}}, 0)(v, \alpha) = (v + e^{i\frac{\pi}{6}}, \alpha)$$

$$(v, \alpha) \mapsto (e^{-i\frac{\pi}{6}}, 0)(v, \alpha) = (v + e^{-i\frac{\pi}{6}}, \alpha)$$

$$(v, \alpha) \mapsto (0, \frac{2\pi}{3})(v, \alpha) = (e^{i\frac{2\pi}{3}}v, \alpha + \frac{2\pi}{3})$$

We take as a fundamental domain for the action of S333 in $\mathbb{C} \times S^1$ the prism of base the hexagon E of Fig. 14 and height $2\pi/3$. The lateral faces of the prism are identified by translation. The top and bottom faces, by translation followed by right rotation of angle $\frac{2\pi}{3}$. *The result, $S333\backslash\mathbb{C} \times S^1$, is an F_1-bundle over S^1 with monodromy of period 3.* The manifold M(S333) is $S333\backslash\mathbb{C} \times S^1$ with the opposite orientation.

Exercise 3. Check that the fiber is a torus.

Exercise 4. Show that M(S632) is an F_1-bundle over S^1 with monodromy of period 6 and that there are coverings $T^3 \to M(S333) \to M(S632)$.

Hint.- S632 is generated by σ, τ and a = $(0, \frac{2\pi}{6})$ (Fig. 15). A fundamental domain for the action of S632 on $E(2) = \mathbb{C} \times S^1$ is the

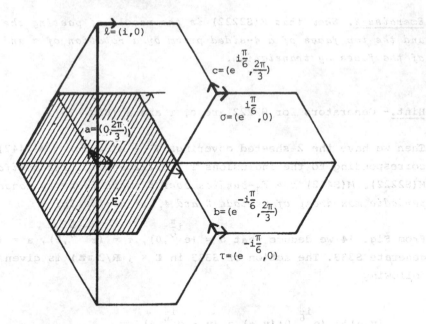

$$\ell = (i,0)$$

$$c = (e^{i\frac{\pi}{6}}, \frac{2\pi}{3})$$

$$\sigma = (e^{i\frac{\pi}{6}}, 0)$$

$$a = (0, \frac{2\pi}{3})$$

$$b = (e^{-i\frac{\pi}{6}}, \frac{2\pi}{3})$$

$$\tau = (e^{-i\frac{\pi}{6}}, 0)$$

Fig. 14. Generators of S333 and hexagon E

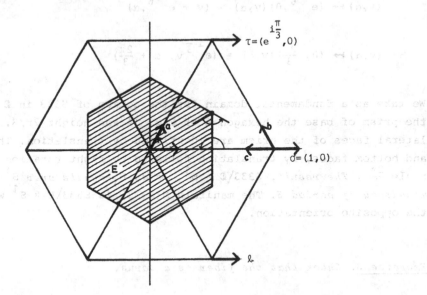

$$\tau = (e^{i\frac{\pi}{3}}, 0)$$

$$\sigma = (1,0)$$

Fig. 15. Generators of S632 and hexagon E

prism of base the hexagon E of Fig. 15 and height $2\pi/6$. Bottom and top faces must be identified by translation followed by right rotation of angle $2\pi/6$.

Thus, we conclude that *there are five manifolds of tessellations in the euclidean plane which correspond to the crystallographic groups (preserving orientation) $T \leqq S2222 \leqq S442$ and $S333 \leqq S632$. All of them are F_1-bundles over S^1 and the monodromies are the identity and periodic homeomorphisms of periods 2, 4, 3 and 6 respectively.*

The relationship of these manifolds to the 3-dimensional euclidean orbifolds is explained in 2.11.

2.8 Involutions in the manifolds of euclidean tessellations

Compare the four tessellations of Fig. 16. All of them have the same symmetry subgroup of $E(2)$, namely T. Nevertheless the last three have *more* symmetry, since the second and the fourth ones have an axis of mirror symmetry and the third and the fourth have axes of glide-reflections (composition of translation along an axis and reflection in the same line). Thus allowing transformations of the euclidean plane which reverse orientation, such as reflections and glide-reflections, there are precisely 17 different possible tessellations, and 17 corresponding (general) crystallographic subgroup of the group of *all* isometries of E^2 (including those reversing orientation). This classification can be seen in Appendix A. The four tessellations T, A, K, M of Figure 16 are those whose subgroups of orientation-preserving symmetries is precisely T (see Table I of Appendix A).

We now see how these observations allow us to define different involutions in the same euclidean manifold of tessellations. Take an axis E, fixed from now on, in the plane E^2 and consider the re- flection r through E. If we take as model tessellation for T the first tessellation of Fig. 16, it is clear that r does not define a map from M(T) into itself because the tessellation under considera- tion does not have any orientation-reversing symmetry. But taking

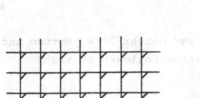

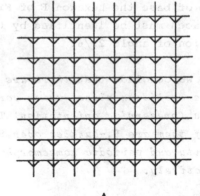

T

A

K

M

Fig. 16. T-tessellations

any of the three other tessellations of Fig. 16, the map r sends
tessellations to tessellations of the same type, thus defining an
orientation-preserving involutive homeomorphism r on M(T). But *note
that this homeomorphism depends on the tessellation selected.* For
instance no tessellation of type K remains fixed under r (for
any choice of r), while (for any choice of r) some tessellation of
type A, resp. type M, will be fixed.

Exercise 1. *Show that r is orientation-preserving on M(T).*

Hint.- Take a disk neighbourhood of a point in E.

What are these involutions and the quotient manifolds? They are certainly involutions of the 3-dimensional torus $T^3 = T^2 \times S^1$; and $T^2 \times 0$ is the set of tessellations which are parallel to E, the involution preserves $T^2 \times 0$, and must be of the form $u \times c$, where u is an orientation-reserving involution and T^2 and c is $z \mapsto \bar{z}$ in $S^1 \subset \mathbb{C}$. The involution $r(A) = u(A) \times c$ has a fixed point set which consists of 4 disjoint circles, while $r(K) = u(K) \times c$ has no fixed points and $r(M) = u(M) \times c$ has two fixed-point circles.

Exercise 2. Prove the last assertion.

Therefore

$$u(A): S^1 \times S^1 \rightarrow S^1 \times S^1$$

is given by

$$(z_1, z_2) \mapsto (\bar{z}_1, z_2) ;$$

$u(K)$ is defined by

$$(z_1, z_2) \mapsto (\bar{z}_1, -z_2) ;$$

and $u(M)$ is given by

$$(z_1, z_2) \mapsto (z_2, z_1)$$

(Fig. 17). The quotients of $T^2 = S^1 \times S^1$ under these involutions are the annulus A, the Klein bottle K, and the Möbius band M, respectively. The quotients of $T^3 = T^2 \times S^1$ under $r(A)$, $r(K)$ and $r(M)$ are shown in Fig. 18, and are, respectively, $S^1 \times S^2$, $(On2|0)$ and $S^1 \times S^2$. The projections

$$p(A): T^3 \rightarrow S^1 \times S^1 ,$$

$$p(K): T^3 \rightarrow (On2|0) \quad \text{and}$$

$$p(M): T^3 \rightarrow S^1 \times S^2$$

are branched coverings in the first and third case, and an un-

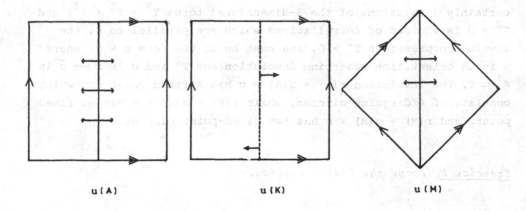

Fig. 17. Involutions on T

branched covering in the second case. The branching sets are shown
in Fig. 18. From 1.7, it follows that (On2|0) is the spherical
tangent bundle of the Klein bottle K. This is not accidental, we
will see in Appendix A that *what we have just obtained are the
spherical tangent bundles of the orbifolds associated with the
tessellations* A, K *and* M. (For some readers acquainted with sections
A.1, A.2 and A.3 of Appendix A, this would be a good place to read
A.4.)

In exactly the same way, in Fig. 19 we have the four tessellations
which have an orientation-reversing symmetry and have the group
S2222 as the orientation-preserving symmetry group. There are four
different orientation-preserving involutions for M(S2222), obtained by reflection in a fixed axis.

Exercise 3. Find the quotient spaces of M(S2222) *under these
involutions. Find also the branching sets of the associated
(branched) coverings.*

Hint.- M(S2222) is a T^2-bundle over S^1,

$$T^2 \times [-1, 1]/(x, -1) \equiv \varphi(x, 1) \ .$$

Its monodromy φ is 180° rotation around 0 (Fig. 20). Just as before,
r is of the form u × c, where u is an involution of T^2 which

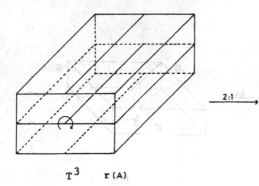

T^3 r(A)

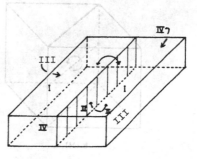

III

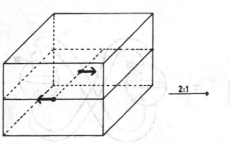

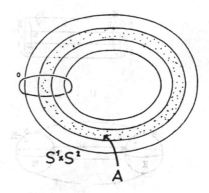

$S^1 \times S^2$

A

r(K)

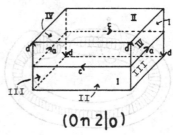

(0 n 2|o)

Figure 18 (continues next page)

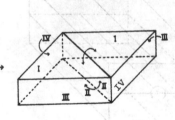

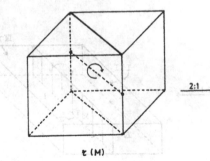

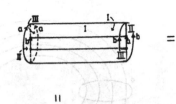

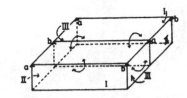

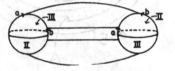

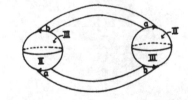

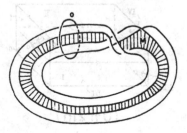

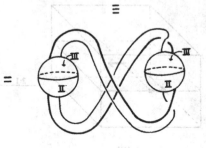

Figure 18

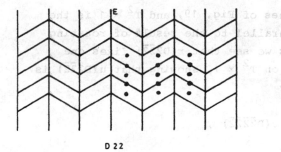

D 22

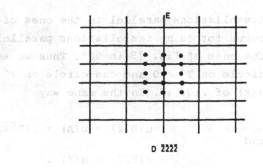

D 2222

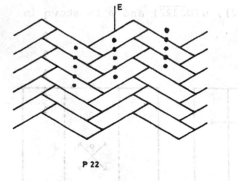

P 22

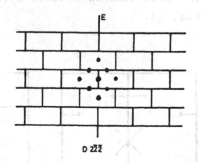

D 2̄2̄2̄

Fig. 19. Four tessellations underlying S2222

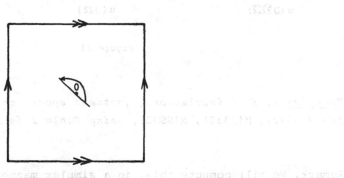

Fig. 20. Monodromy for M(S2222)

commutes with the reflection through O, and

$$c : [-1, 1] \to [-1, 1]$$

sends t to -t. Then $T^2 \times 0 \subset M(S2222)$ is the torus formed by tessellations parallel to the ones of Fig. 19, and $T^2 \times 1$ is the torus formed by tessellations parallel to the result of rotating the ones of Fig. 19 in 90°. Thus we see that $r(D2\overline{2}\overline{2})$ fixes one circle on $T^2 \times 0$ and one circle on $T^2 \times 1$, showing that $u(D2\overline{2}\overline{2})$ is $u(M)$ of Fig. 17. In the same way

$$u(D22) = u(A) = u(D\overline{2}\overline{2}\overline{2}\overline{2}) \ ,$$

and

$$u(P22) = u(K) \ .$$

The relationship between $u(P22)$, $u(D22)$, $u(D\overline{2}\overline{2}\overline{2}\overline{2})$ and φ is shown in Fig. 21.

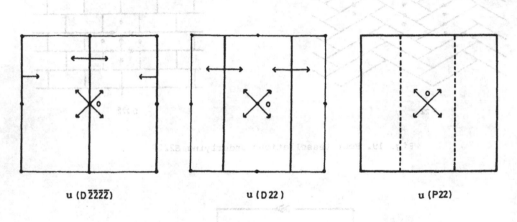

u (D$\overline{2}\overline{2}\overline{2}\overline{2}$)   u (D22)   u (P22)

Figure 21

Exercise 4. *Find involutions, quotient spaces and branching sets for* M(S442), M(S333), M(S632), *using Table I in Appendix A.*

Remark. We will compute this, in a simpler manner, in 4.6.

2.9 The fundamental groups of the manifolds of euclidean tessellations

Let Γ be a crystallographic orientation-preserving subgroup of $E(2)$ and consider $\tilde{\Gamma} = p^{-1}\Gamma$, where $p:\tilde{E}(2) \to E(2)$ is the universal covering of $E(2)$ (see section 2). Since the sequence

$$0 \to \mathbb{Z} \longrightarrow \tilde{E}(2) \xrightarrow{p} E(2) \to 0$$

$$1 \mapsto (0, 2\pi)$$

is exact, we have the short exact sequence

$$0 \to \mathbb{Z} \to \tilde{\Gamma} \xrightarrow{p} \Gamma \to 0 \quad . \tag{1}$$

Thus $\tilde{E}(2)/\tilde{\Gamma}$ is diffeomorphic to $E(2)/\Gamma$. On the other hand $\tilde{\Gamma}$ acts freely (i.e. without fixed points) in $E(2)$, because if $(w, r) \in \tilde{\Gamma}$ and $(v, t)(w, r) = (v, t)$ then

$$(v + e^{it}w, t+r) = (v, t)$$

is in $\mathbb{C} \times \mathbb{R}$; therefore $(w, r) = (0, 0)$. *Thus the fundamental group of the manifold of tessellations defined by Γ is $\tilde{\Gamma}$.*

Exercise 1. *Show that the extension* (1) *is central.*

Solution.- Let $(v, t) \mapsto (v, t)$ be an injection (not a monomorphism) of Γ in $\tilde{\Gamma}$. We have

$$(0, 2\pi m)(v, t)(0, -2\pi m) = (ve^{i2\pi m}, t) = (v, t)$$

(for group extensions consult [Rot]).

The classical construction of the universal cover of $E(2)$ considers the points of $\tilde{E}(2)$ as classes of "paths of isometries" of E^2 connecting the identity isometry $(0, 0)$ with some other (v, α) (Fig. 22). It is always possible to normalize the path by first taking a path of rotations with angles $t(2\pi n + \alpha)$, $0 \le t \le 1$, connecting $(0, 0)$ to $(0, \alpha)$ and then taking a path of translations of vectors tv, $0 \le t \le 1$, connecting $(0, \alpha)$ to (v, α). A convenient way of representing this particular element of $\tilde{E}(2)$ is to endow the pointer (v, α) with the integer n: (v, α, n). In this notation, $(0, 0, n)$ is an element of the kernel of

$$p:\tilde{E}(2) \to E(2) \quad .$$

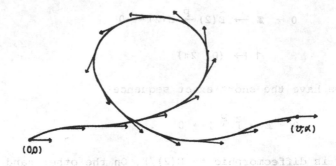

$(\widetilde{v}, \alpha)$

$(0,0)$

Fig. 22. An element of $\widetilde{E}(2)$

Thus, the elements of $\widetilde{\Gamma}$ are the (v, α, n)'s where $(v, \alpha) \in \Gamma$. Thus, an element of

$$\widetilde{\Gamma} = \pi_1(M(\Gamma), c)$$

is a *path of tessellations* which starts and ends at the base tessellation c (Fig. 23); for example, a path of rotations with angles $(2\pi n + \frac{2\pi}{m})t$, $0 \leq t \leq 1$ about a center of symmetry of order m is an element of $\widetilde{\Gamma}$. Thus, for instance, $\widetilde{\gamma}^6 = (0, 0, -1)$, for $\widetilde{\gamma} = (0, -\frac{\pi}{3}, 0)$ of Fig. 23. It is very useful to think of the elements of $\pi_1(M(\Gamma), c)$ as "paths of rotations" of the tessellation c, whose final rotation leaves c invariant ("rotation" is used in the wide sense so as to include rotations of infinite radius, i.e. translations). A path of rotations which performs n complete turns represents the element n of the center of

$$\pi_1(M(\Gamma), c) = \widetilde{\Gamma} \quad .$$

Of course, the image of $\widetilde{\gamma} \in \widetilde{\Gamma}$ by $p: \widetilde{\Gamma} \to \Gamma$ is just the final rotation of the path of rotations representing $\widetilde{\gamma}$.

2.10 Presentations of the fundamental groups of the manifolds $M(\Gamma)$

The group S2222 has the presentation

$$|a, b, c, d: a^2 = b^2 = c^2 = d^2 = abcd = 1|$$

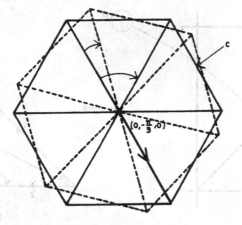

$(0, -\frac{\pi}{3}, 0)$

Fig. 23. The element of $\tilde{\gamma} = (0, -\frac{\pi}{3}, 0)$, for $\tilde{\Gamma} = \pi_1(M(S632),c)$ is a path of rotations of the wheel

and the groups $S\ell mn$ *have the presentations*

$$|a, b, c: a^\ell = b^m = c^n = abc = 1|$$

for

$$(\ell, m, n) = \begin{cases} 442 \\ 632 \\ 333 \end{cases}$$

<u>Proof</u>.- The analysis made in section 7 shows that the generators are correct and that they satisfy the given relations. It only remains to be seen that no more relations are needed. Let us see the case for $\Gamma = S442$. This group acts freely on $\mathbb{C}-V$, where V is the set of vertices of $c(S442)$, hence there is a regular covering

$$f: \mathbb{C}-V \rightarrow S_o^2 \quad,$$

where S_o^2 is the 2-sphere minus three points (see Appendix A). Hence S442 is isomorphic to the quotient of $\pi_1(S_o^2, 0)$ under the subgroup $f_*(\pi_1(\mathbb{C}-V, \tilde{o}))$ (Fig. 24). We take the presentation

$$|a, b, c: abc = 1|$$

for $\pi_1(S_o^2, 0)$. As generators for $\pi_1(\mathbb{C}-V, \tilde{o})$, we take paths starting

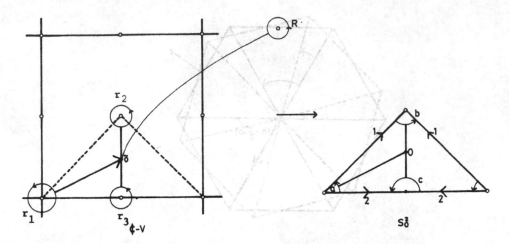

Fig. 24. The covering $\mathbb{C}-V \to S_o^2$

at the base point $\tilde{O}$ and surrounding the vertices V (Fig. 24). The images of r_1, r_2, r_3 under f_* are a^4, b^4, c^2, respectively. The image of any other generator, for instance R, of $\pi_1(\mathbb{C}-V, \tilde{O})$ is an element conjugated to a^4, b^4 or c^2. Hence

$$S442 = \pi_1(S_o^2, \tilde{O})/f_*(\pi_1(\mathbb{C}-V, \tilde{O})) = |a, b, c: a^4 = b^4 = c^2 = abc = 1|.$$

To find a presentation for $\tilde{\Gamma}$ we consider the short exact sequence (see [Z])

$$0 \to \mathbb{Z} \xrightarrow{i} \tilde{\Gamma} \xrightarrow{p} \Gamma \to 0$$

Let $\varphi : \Gamma \to \tilde{\Gamma}$ be the injection (as a *set*) map

$$\varphi(v, t) = (v, t, 0) .$$

Let $i(1) = z$. Then, if

$$|g_1, \ldots, g_s : r_1, \ldots, r_t|$$

is a presentation of Γ, the elements $\{z, \varphi(g_1), \ldots, \varphi(g_s)\}$ generate $\tilde{\Gamma}$. To find the relations note that

$$r_j(\varphi(g_1), \ldots, \varphi(g_s)) = \tilde{r}_j$$

is in the kernel of p, hence $\tilde{r}_j = z^{n_j}$, for some $n_j \in \mathbb{Z}$. We claim that

$$|z, \varphi(g_1), \ldots, \varphi(g_s) : [z, \varphi(g_k)] = 1, \ k = 1, \ldots, s; \ \tilde{r}_j = z^{n_j}, \ j = 1, \ldots, t|$$

presents $\tilde{\Gamma}$. In fact, any relation is some word of $\tilde{\Gamma}_i$, and since

$$[z, \varphi(g_k)] = 1 ,$$

it can be written as

$$z^n w(\varphi(g_1), \ldots, \varphi(g_s)) = 1 .$$

Then $w(g_1, \ldots, g_s) = 1$, hence w depends on $r_1, \ldots, r_t$. Hence $z^n w = 1$, depends on

$$\tilde{r}_j z^{-n_j} = 1, \ j = 1, \ldots, t .$$

To write a presentation for $\tilde{\Gamma}$, it is enough to write $\tilde{r}_j$ as a power of z. For instance, take

$$S442 = |a, b, c : a^4 = b^4 = c^2 = abc = 1| .$$

We have

$$\varphi(a) = \varphi(0, \frac{\pi}{2}) = (0, \frac{\pi}{2}, 0) ,$$

$$\varphi(b) = (1, \frac{\pi}{2}, 0) ,$$

$$\varphi(c) = (1, \pi, 0).$$

Hence

$$\varphi(a)^4 = (0, 0, 1) = z ,$$

$$\varphi(b)^4 = (0, 0, 1) = z ,$$

$$\varphi(c)^2 = (0, 0, 1) = z ,$$

$$\varphi(a)\varphi(b)\varphi(c) = (0, 0, 1) .$$

Hence

$$\tilde{S}442 = |z, A, B, C : A^4 = B^4 = C^2 = ABC = z| = |A, B, C : A^4 = B^4 = C^2 = ABC| ,$$

where $A = \varphi(a)$, $B = \varphi(b)$, $C = \varphi(c)$.

Exercise 1. _The same happens for the other crystallographic groups,_
i.e. the presentation for $\tilde{\Gamma}$ _comes from the presentation for_ Γ
"deleting the 1".

2.11 The groups $\tilde{\Gamma}$ as 3-dimensional crystallographic groups

The group $\Gamma \le E(2)$ has been interpreted as a group acting by
diffeomorphisms on $ST(E^2)$, and the manifold $M(\Gamma)$ is just the quotient $ST(E^2)/\Gamma$ under this action. We have also seen in section 9
that $ST(E^2)/\Gamma$ is also $\tilde{ST}(E^2)/\tilde{\Gamma}$.Thus the manifold $M(\Gamma)$ is the quotient of $E^3 \cong \tilde{ST}(E^2)$ under the action of the group $\tilde{\Gamma}$. We claim
that $\tilde{\Gamma}$ _is a subgroup of the group of isometries_ $E(3)$ _of the euclidean_
space E^3, i.e. _that_ $\tilde{\Gamma}$ _acts as a 3-dimensional crystallographic_
group of the euclidean space E^3. Therefore the manifolds $M(\Gamma)$ are
euclidean 3-dimensional orbifolds (see Appendix A). For instance,
the action of Γ = S442 on $\mathbb{C} \times (\mathbb{R}/2\pi \mathbb{Z})$ is generated by the translations

$$(v, \alpha) \mapsto (v + 1, \alpha)$$

and

$$(v, \alpha) \mapsto (v + i, \alpha) ,$$

and by the composition of a translation

$$(v, \alpha) \mapsto (v, \alpha + \frac{\pi}{2})$$

and a rotation

$$(v, \alpha + \frac{\pi}{2}) \mapsto (ve^{i\frac{\pi}{2}}, \alpha + \frac{\pi}{2}) .$$

Exactly the same happens with the other groups.

This raises a very interesting question. What are the crystallographic subgroups Λ of $E(3)$ such that E^3/Λ is an orientable 3-
manifold, and what are these 3-manifolds? It turns out to be that
they are the five just described in section 7, and the spherical
tangent bundle of the orbifold P22 (see 4.8)! This is more amazing
when one learns that there are precisely 219 pairwise non-isomorphic

crystallographic subgroups of E(3). The reader interested in understanding this and learning what orbifolds E^3/Λ are possible, should read the beautiful paper [BS], and the thesis of Dumbar [Du1].

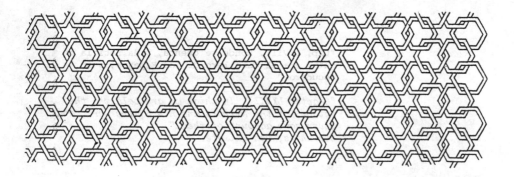

Appendix A: Orbifolds

"Cáp. LXVIII, Donde se cuenta, y da noticia de quién era el Caballero de los Espejos, y su escudero."

"Chapter LXVIII, In which we are told who the Knight of the Mirrors and his Squire were, and given some account of them."
Cervantes, Don Quixote

A.1. Introduction

The reader acquainted with orbifolds could well omit the following and start with section A.4.

Given a closed n-manifold M it is possible to construct the universal cover $\tilde{M}$ of M (see [ST]). The points of $\tilde{M}$ are classes of paths starting at some base point b $\in$ M; two paths finishing in the same point are equivalent if they are homotopic while keeping their end points fixed. Thus $\tilde{M}$ is just the set of the "different ways of reaching the points of M". Sometimes it is possible to fix a particular representative path in each class defining a point of $\tilde{M}$. This happens, for instance, with some Riemannian manifolds M of non-positive sectional curvature where we can represent a class by a geodesic. In that case $\tilde{M}$ has the following interpretation (due to Thurston [Th1]). Imagine you are at b, and look around; what you see is $\tilde{M}$, assuming M is illuminated, because each point c of M emits light and this light reaches you after travelling through all possible geodesics. Thus you see as many different images of c as there are different geodesics. To make this more concrete, think of the torus M endowed with a Riemannian metric of zero curvature (Fig. 1). The universal covering is the euclidean plane (Fig. 1).

Together with $\tilde{M}$ there is the group of covering transformations (isomorphic to $\pi_1(M)$) and we can think of M as the quotient of $\tilde{M}$ under

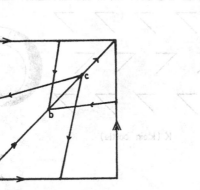

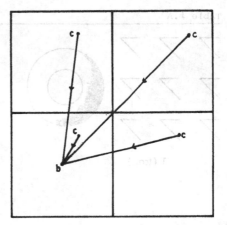

Fig. 1. Torus   Fig. 2. Euclidean plane

the action of this group. This group acts freely and properly-
discontinuously on $\tilde{M}$, as in the case of the torus in which the group
(isomorphic to $\mathbb{Z}^2$) is generated by two independent translations.
*Thus M embodies $\tilde{M}$ and a group acting freely and properly-discontinu-
ously on $\tilde{M}$.* Extensive work has been done on groups G acting properly-
discontinuously (but not necessarily freely) in n-manifolds $\tilde{M}$.
Satake and Thurston understood that the quotient "orbifold" M is a
good device for embodying the information given by $\tilde{M}$ and G, gener-
alizing in this way the concept of manifold. An orbifold then should
be understood as a sort of kaleidoscope in which you can see the
"universal cover" $\tilde{M}$. As an example consider the *closed, euclidean
2-orbifolds*. These are the quotients of the euclidean plane E^2 under
the action of the 17 crystallographic groups (i.e. subgroups Γ of
the group of isometries of E^2 which act properly-discontinuously and
such that E^2/Γ is compact); these quotients (Table I and Plate I)
are 2-manifolds with boundary, and they are endowed with enough
information to reconstruct the "universal cover", i.e. E^2, and the
corresponding crystallographic group. The four first orbifolds in
the table are the torus T, the Klein bottle K, the annulus A and the
Möbius band M. The torus and Klein bottle are understood here just
as 2-manifolds and their universal covers are shown in the right
hand column of the table. The crystallographic group for T is
generated by two independent translations, and for K, by one translation and one glide-reflection. The case of A (and of M) is already
different. Their usual universal cover (thinking of A and M just as
2-manifolds with boundary) is an infinite strip and the group is

Table I.A

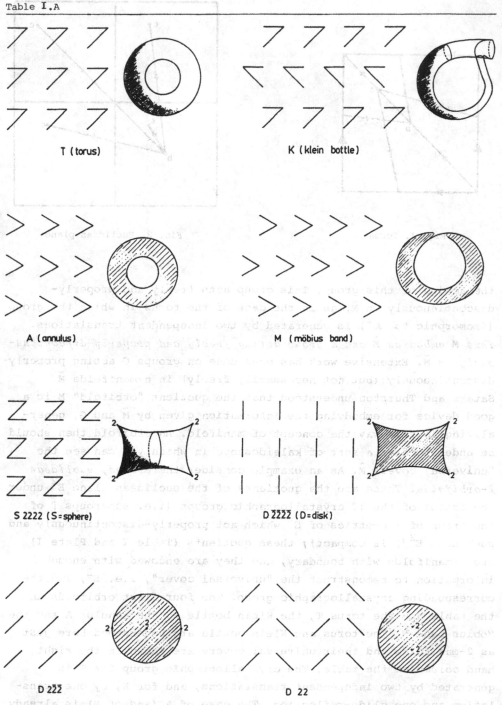

T (torus)

K (klein bottle)

A (annulus)

M (möbius band)

S 2222 (S=sphere)

D 2222 (D=disk)

D 222

D 22

Table **I**.A (continued)

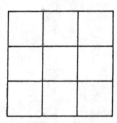 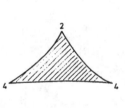 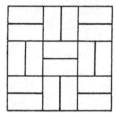

P 22 (P=projective)

S 442

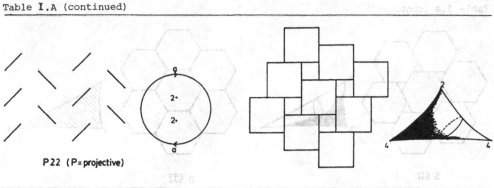

D $\overline{442}$

D $4\overline{2}$

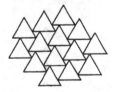   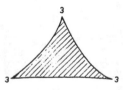

S 333

D $\overline{333}$

D $3\overline{3}$

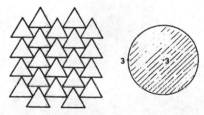

Table I.A (continued)

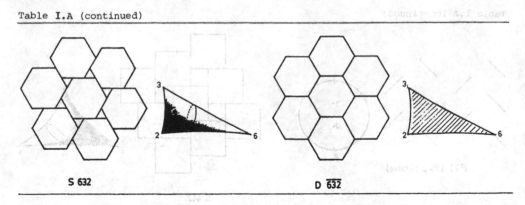

S 632   D $\overline{632}$

Table I.A.- The 17 closed, euclidean 2-orbifolds (right) with their universal
covering (left), and their notation (under)

generated by a translation in the case of A and by a glide-
reflection in the case of M. But here we think of A and M as *closed*
orbifolds, and this additional information is given in the form of
a marked ("silvered") boundary. The meaning of this is that the infinite strip should be reflected in its silvered boundary. The resulting universal covering is E^2, and the crystallographic group
has an additional generator, which is a reflection in a line. The
fifth orbifold has a new feature; namely, four points marked with
the number 2. The meaning of this is the following. We have an
atlas of charts for S^2 modelled on $\mathbb{R}^2$, but a chart U, covering one
of the four marked points x, should be considered together with the
2-fold covering $\tilde{U}$ of U branched over the marked point x (Fig. 3).

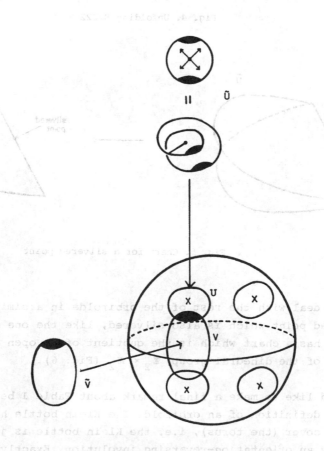

Fig. 3. Atlas for S2222

Thus, charts parametrizing marked points are "folded" charts, and
to obtain the universal cover we just think of them as "unfolded
charts" (Fig. 4). With this point of view, a silvered point has a
chart as shown in Fig. 5.

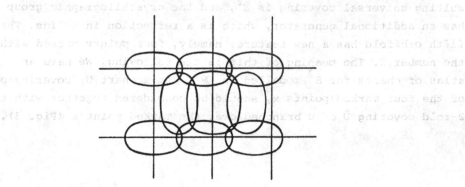

Fig. 4. Unfolding S2222

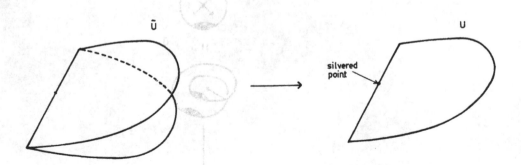

Fig. 5. Chart for a silvered point

We can deal with the rest of the orbifolds in a similar way. Thus,
a marked point which is also silvered, like the one marked $\bar{2}$ in
$D\bar{2}\bar{2}\bar{2}\bar{2}$, has a chart which is the quotient of an open disk by the
action of the dihedral group $\mathbb{Z}_2 \times \mathbb{Z}_2$ (Fig. 6).

I would like to make a final remark about Table I before giving the
formal definition of an orbifold. The Klein bottle has an orientable
2-fold cover (the torus), i.e. the Klein bottle is just the quotient
of T by an orientation-reversing involution. Exactly the same occurs

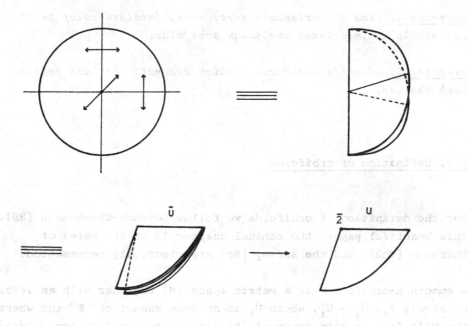

Fig. 6. Chart for a marked silvered point

with the "non-orientable" orbifolds, i.e. those with silvered boundary or those whose underlying 2-manifold is non-orientable, such as
M or P22. Thus, the reader should check that the 5 orientable orbifolds T, S2222, S442, S333, S632 have enough orientation-reversing
involutions to create the 12 remaining non-orientable orbifolds
(Fig. 7). This remark would serve to show that our intuitions about
manifolds have a wider range of applications, making orbifolds useful objects.

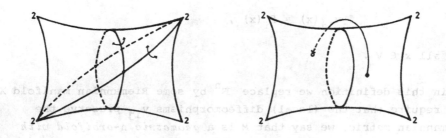

Fig. 7. Some involutions in S2222

Exercise 1. *Find all orientation-reversing involutions of the 5 orientable 2-dimensional euclidean orbifolds.*

Exercise 2. *Identify the mosaics which decorate the last page of each chapter.*

A.2. Definition of orbifolds

For the definition of orbifolds we follow Bonahon-Siebenmann [BS]. This beautiful paper, the seminal chapter 13 of the notes of Thurston [Th2], and the survey [Sc] are vigorously recommended.

A smooth n-orbifold M is a metric space $|M|$ together with an *atlas* of *charts* $f_i: \tilde{U}_i \to U_i$, where $\tilde{U}_i$ is an open subset of $\mathbb{R}^n$ and where the U_i's form an open cover of $|M|$, and where the f_i's are *folding maps*, i.e. the group G_i of diffeomorphisms of $\tilde{U}_i$ preserving fibers of f_i must be finite and the induced map $\tilde{U}_i/G_i \to U_i$ must be a homeomorphism. The charts are *compatible*, meaning that, whenever

$$f_i(x_i) = f_j(x_j) ,$$

there are open neighbourhoods $\tilde{V}_i$, $\tilde{V}_j$ of x_i, x_j in $\tilde{U}_i$, $\tilde{U}_j$, and a diffeomorphism

$$v_{ij} : \tilde{V}_i \to \tilde{V}_j$$

such that

$$f_j v_{ij}(x) = f_i(x) ,$$

for all $x \in \tilde{V}_i$.

If in this definition we replace $\mathbb{R}^n$ by some Riemannian manifold X and require that the (local) diffeomorphisms v_{ij} preserve the Riemannian metric, we say that M is a *geometric n-orbifold with model X*. If X is the euclidean, spherical or hyperbolic n-space, E^n, S^n or H^n, the orbifold will be called *euclidean*, *spherical* or *hyperbolic*.

Each point x of |M| has an associated *isotropy group* G_x. This is
the subgroup of automorphisms of $\mathbb{R}^n = T_{x_i} \tilde{U}_i$ induced by the elements
of the stabilizer of x_i in G_i, where $f_i : \tilde{U}_i \to U_i$ is a chart such
that $f_i(x_i) = x$. The group G_x is well defined up to conjugacy in
$GL(\mathbb{R}, n)$. When M has model the Riemannian manifold X, G_x is a subgroup of $O(n)$.

The definition of geometric n-orbifold with model X contains as a
particular case the concept of *geometric n-manifold with model* X
when the groups G_i all reduce to the identity element. We thus
think of an orbifold as a set of folded charts (folded by groups G_i
preserving some prescribed "structure") such that the change from
one chart to another overlapping one preserves "structure" locally.
Two atlases define the same orbifold if their union is again a
compatible atlas. An *isomorphism* of orbifolds with model X is a
homeomorphism sending one atlas to a compatible atlas. An orbifold
with model X is *oriented* if the local homeomorphisms in the construction of its atlas preserve orientation.

A.3. The 2-dimensional orbifolds

We have given examples of orbifolds in the first section. All of
them were quotients of E^2 under subgroups of the group of euclidean
isometries of E^2. Hence the *orbifolds of Table I are euclidean orbifolds*. In general given X and a group H of isometries acting
properly-discontinuously on M, the quotient M/H is an orbifold with
model X. These are the *good* orbifolds in the sense of Thurston, i.e.
orbifolds covered by manifolds. There are also *bad* (i.e. not good)
orbifolds. For instance, the orbifolds of Fig. 8 cannot be unfolded
to manifolds, because in any unfolding of S3 (Fig. 9) the disk U_1
would be a number of disks covering it 3 to 1 and the disk U_2 would
be covered 1 to 1, which is impossible, since they have the same
boundary curve a.

Exercise 1. Prove that Saβ, Dᾱ *and* Dᾱβ̄ *are bad.*

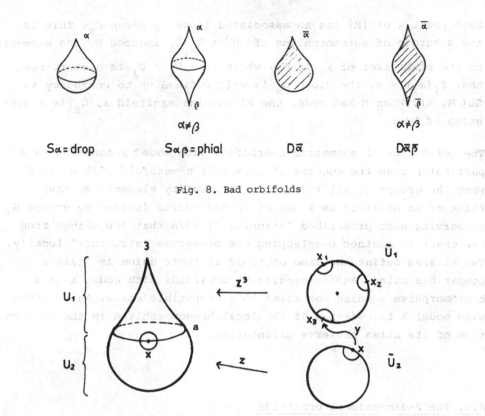

$\alpha \neq \beta$ $\alpha \neq \beta$

$S\alpha = $ drop $S\alpha\beta = $ phial $D\bar{\alpha}$ $D\bar{\alpha}\bar{\beta}$

Fig. 8. Bad orbifolds

Fig. 9. The orbifold S3

Examples of spherical orbifolds are shown in Table II (see also Plate II). They are the quotients of S^2 under the finite subgroups of O(3). The non-orientable orbifolds are the quotient of the orientable ones (Smm, Sm22, S332, S432, S532) under orientation-reversing involutions.

Exercise 2. Find all orientation-reversing involutions of the orientable spherical 2-orbifolds.

It turns out to be that *all good 2-orbifolds can be provided with euclidean, spherical or hyperbolic atlases.* This is a result of Thurston who also introduced the Euler characteristic of an orbifold M, as follows:

Table II.A

S 332 $x=\frac{1}{6}$

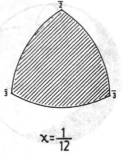

D $\bar{3}\bar{3}\bar{2}$ $x=\frac{1}{12}$

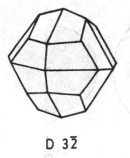

D $3\bar{2}$ $x=\frac{1}{12}$

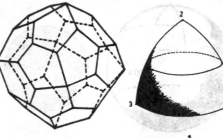

D 432 $x=\frac{1}{12}$

D $\bar{4}3\bar{2}$ $x=\frac{1}{24}$

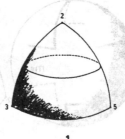

S 532 $\bar{x}=\frac{1}{30}$

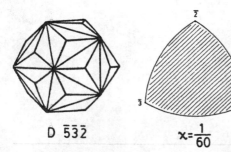

D $\bar{5}3\bar{2}$ $x=\frac{1}{60}$

Table II.A (continued)

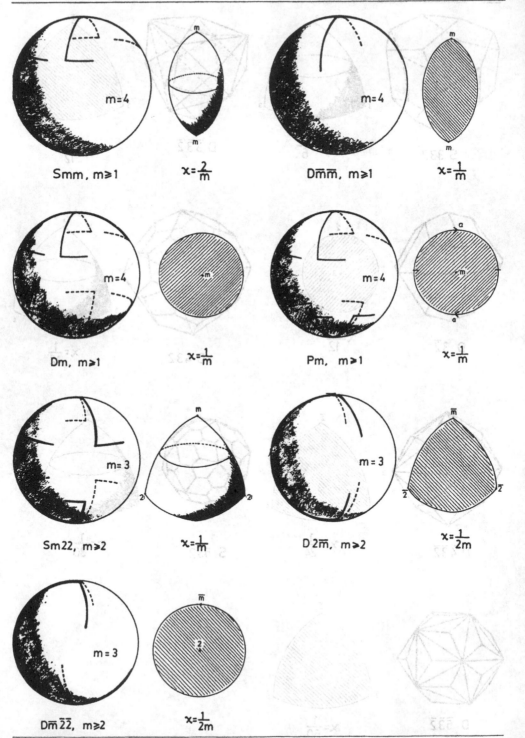

Smm, m≥1 $x=\frac{2}{m}$ Dm̄m̄, m≥1 $x=\frac{1}{m}$

Dm, m≥1 $x=\frac{1}{m}$ Pm, m≥1 $x=\frac{1}{m}$

Sm22, m≥2 $x=\frac{1}{m}$ D2m̄, m≥2 $x=\frac{1}{2m}$

Dm̄2̄2̄, m≥2 $x=\frac{1}{2m}$

$$\chi(M) = \sum_{C_i} (-1)^{\dim C_i} (\text{order } G(C_i))^{-1}$$

where C_i ranges over the cells of a cellular decomposition of $|M|$,
such that the points of the interior of a cell C_i all have the same
isotropy group $G(C_i)$. The last column of Table II contains the
Euler characteristics of the corresponding orbifolds. Notice that
$\chi(M)$ is zero for the euclidean orbifolds (and > 0 for the spherical
orbifolds).

Then Thurston shows in [Th2] that *the good, connected, closed 2-
orbifolds with Euler characteristic > 0, 0, < 0 are respectively
spherical, euclidean, or hyperbolic.* The only possibilities for
isotropy groups in the case of 2-orbifolds are finite subgroups of
$O(2)$, i.e. cyclic groups $\mathbb{Z}_m$, which are subgroups of $SO(2)$, and
dihedral groups D_{2m} which contain an orientation-preserving subgroup
$\mathbb{Z}_m$. Hence an underlying metric space $|M|$ of a 2-orbifold must be
some compact 2-manifold with silvered boundary (isotropy group D_2)
containing a finite number of "corner points" with isotropy group
D_{2m}, and whose interior contains a finite number of "cone points"
with isotropy group $\mathbb{Z}_m$. The notation used in Tables I and II is a
symbol for the manifold $|M|$ followed by the orders of the isotropy
groups of the cone-points and half the orders of the isotropy groups
of the corner points, marked with a hyphen. It is an easy exercise
to show that there are no more orbifolds with $\chi(M) \geq 0$ than the
ones listed in Tables I and II. It remains to be shown that the rest
of the orbifolds (with $\chi(M) < 0$) are hyperbolic and Thurston does
this by direct construction (see [Th2], for details).

A.4. The tangent bundle

The main topic of the book are the manifolds of tessellations. Up
to orientation we can think of a manifold of tessellations as the
set of orbits of the action of a group Γ of isometries of X (= S^2,
E^2 or H^2) on $ST(X)$. This quotient space $\Gamma \backslash ST(X)$ has a different
interpretation. In the context of orbifolds, $\Gamma \backslash ST(X)$ is just the
spherical tangent bundle of the orbifold $\Gamma \backslash X$.

Given an n-orbifold M we can define *the tangent bundle* TM just as
for manifolds. Assume

$$\{f_i : \tilde{U}_i \to U_i ; \ i \in I\}$$

is an atlas for M. Form the atlas

$$\{g_i : \tilde{U}_i \times \mathbb{R}^n \to (\tilde{U}_i \times \mathbb{R}^n)/G_i\}$$

where $g \in G_i$ acts on $\tilde{U}_i \times \mathbb{R}^n$ by

$$(x_i, w) \mapsto (gx_i, dg_{x_i} w) \ .$$

We must construct $|TM|$. This is done by identifying the spaces
$\{\tilde{U}_i \times \mathbb{R}^n/G_i\}$: $g_i(x_i, w)$ is identified with $g_j(x_j, (dv_{ij})_{x_i} w)$. Thus
TM is a 2n-orbifold and there is a continuous map

$$p: |TM| \to |M|$$

sending $g_i(x_i, w)$ to $f_i(x_i)$.

The tangent bundle is a particular case of a more general concept.
A *bundle* M → N with total space the orbifold M, *base* the orbifold
N and *fiber* the manifold F is a continuous map

$$p: |M| \to |N|$$

such that, for each $x \in |N|$ there are charts

$$f_i : \tilde{U}_i \to U_i \ \ , \text{ with } x \in U_i,$$

and

$$g_i : \tilde{U}_i \times F \to p^{-1}(U_i)$$

such that the following diagram is commutative

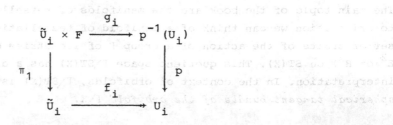

Moreover the folding group of g_i respects the product structure
so that the fiber $p^{-1}(x) \subset M$ is an orbifold quotient of F.

If the orbifold M is good, i.e. M is the quotient of a manifold X
under the action of a group Γ acting properly-discontinuously on X,
the tangent bundle TM is the orbifold $\Gamma \backslash TX$, where $g \in \Gamma$ acts on TX
by

$$(x, w) \mapsto (gx, dg_x w) .$$

Thus, *the manifolds of tessellations studied in this book are, up
to orientation, the spherical tangent bundles of 2-dimensional
oriented orbifolds.*

There is a common feature among these bundles. We have seen in Chapter I that
the tangent bundles of the surfaces F_g, or N_k are, respectively, $(Oog|-\chi(F_g))$
or $(Onk|-\chi(N_k))$. The reader can check that the *rational Euler number* e_0 of the
tangent space of an oriented 2-orbifold equals the rational Euler number of the
base 2-orbifold (see 4.4). For instance, the orbifold S2222 with Euler number
$\chi(S2222) = 0$ has tangent bundle

$$-(Oo0|-2; \ (2,1), \ (2,1), \ (2,1), \ (2,1)) = (Oo0|-2; \ (2,1), \ (2,1), \ (2,1), \ (2,1))$$

$$= (0; \ 0; \ 1/2, \ 1/2, \ 1/2, \ 1/2)$$

with rational Euler number zero (see 4.4 for the notation used here).

Let c be a tessellation of $X = E^2$, S^2 or H^2 which has Γ as full
group of symmetries and Γ_0 as index 2 subgroup of orientation-
preserving isometries fixing c. If r is reflection in an axis E of
X, we have considered in 2.8 (and we will consider it again in 3.15
and 5.6) the involution

$$r(c) : M(c) \to M(c)$$

$$t \mapsto r(t)$$

which sends the tessellation t (of type c) into r(t). We want to
show that the quotient M(c)/r(c) is, up to orientation, the spherical
tangent bundle of the orbifold $\Gamma \backslash X$.

To see this, let $\gamma \in \Gamma$ be an orientation-reversing isometry fixing
c, and note that there must be an orientation-preserving isometry
of X, λ, such that $r = \gamma\lambda$. If $Iso^+(X)$ is the group of orientation-

Notes to Plate I: see page 224

Notes to Plate II: see page 230

preserving isometries of X, the map $r(c)$ induces on $\text{Iso}^+(X)/\Gamma_0$ (which is orientation-preserving diffeomorphic to $M(c)$) the involution

$$g\Gamma_0 \mapsto g\Gamma_0(c) = g(c) \mapsto rg(c) = \gamma\lambda g\gamma(c) \mapsto \gamma\lambda g\gamma\Gamma_0 \ .$$

This involution induces, under the map $g \mapsto g^{-1}$, the involution

$$\Gamma_0 \backslash \text{Iso}^+(X) \rightarrow \Gamma_0 \backslash \text{Iso}^+(X)$$

$$\Gamma_0 g \mapsto \Gamma_0 \gamma g\lambda^{-1}\gamma \ .$$

Under the identification

$$\text{Iso}^+(X) \rightarrow \text{ST}(X) \ ,$$

$$g \mapsto g(b)$$

where we select the element b to be invariant under r, the last involution becomes

$$\Gamma_0 \backslash \text{ST}(X) \rightarrow \Gamma_0 \backslash \text{ST}(X)$$

$$\Gamma_0[g(b)] \mapsto \Gamma_0 \gamma g\lambda^{-1}\gamma(b) = \Gamma_0 \gamma[g(b)] = \gamma\Gamma_0[g(b)].$$

which is the involution induced by left multiplication by γ. Thus $M(c)/r(c)$ *is orientation-reversing diffeomorphic to* $\Gamma\backslash\text{ST}(X)$, *which is the spherical tangent bundle of* $\Gamma\backslash X$.

Chapter Three: Manifolds of Spherical Tessellations

"Cáp. CII, De lo que le sucedió a Sancho Panza
rondando su Insula."

*"Chapter CII, What happened to Sancho Panza on the
Rounds of his Isle."*
Cervantes, Don Quixote

The spherical tessellations are represented by the platonic solids
or regular polyhedra. Thus the set of positions of a platonic solid
inscribed in the 2-sphere S^2 is a closed 3-manifold. The fundamental
group of such a manifold is an extension of $\mathbb{Z}_2$ by the group of
isometries of the platonic solid. The subgroup $\mathbb{Z}_2$ is the center of
the group. The manifold can be thought of as the spherical tangent
bundle of a 2-dimensional spherical orbifold, or as a 3-dimensional
spherical orbifold, i.e. the quotient of S^3 under a finite subgroup
of SO(4). Some material of this chapter was borrowed in part from
[DuV].

3.1 The isometries of the 2-sphere

The group of isometries of the euclidean space E^3 which fix the
origin and preserve orientation is SO(3), i.e. the group of orthogonal transformations with determinant +1. This group SO(3) is a
Lie group which acts on S^2 (and, by the differential, also in
$ST(S^2)$) transitively and *effectively* (no element of SO(3), different
from the identity, fixes S^2 pointwise). Taking a base-pointer b of
$ST(S^2)$ we have the diffeomorphism

$$SO(3) \rightarrow ST(S^2)$$

$$\varphi \mapsto d\varphi(b)$$

By means of this diffeomorphism we identify SO(3) with $ST(S^2)$, (Fig. 1.1).

3.2 The fundamental group of SO(3)

We already know (1.1) that

$$\pi_1(SO(3)) \cong \mathbb{Z}_2 .$$

We illustrate this with the "experiment of the belt" that I learned in a lecture of L. Siebenmann. In Fig. 1a we see how, after moving the belt in the way indicated, it undergoes two complete turns. Fig. 1b is another version of this situation, also due to L. Siebenmann.

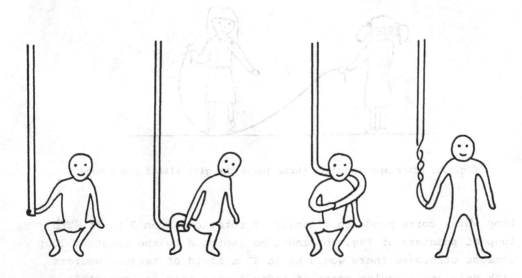

Fig. 1a. Holding the belt, it recovers its original position but with two complete turns

The explanation of the experiment is the following. The belt is the graph of a loop of pointers in SO(3) (Fig. 2). The isotopy of Fig. 1 takes place in SO(3) $\times$ S^1. The projection of this isotopy onto $ST(S^2)$ gives a homotopy between the constant loop of pointers and a

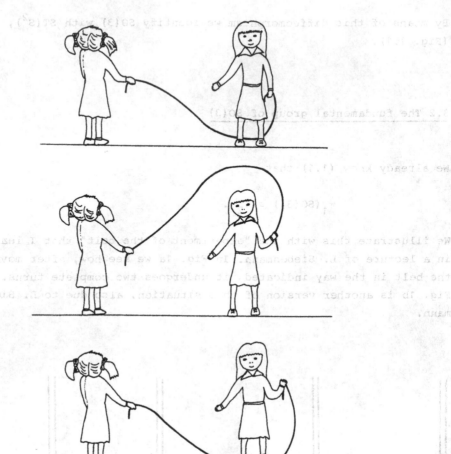

Fig. 1b. After approximately three jumps the girl kisses the ground

loop which corresponds to a family of rotations from 0 to 4π. The loop of pointers of Fig. 2b cannot be isotoped to the constant loop because otherwise there would be in S^2 a field of tangent vectors with only one singular point of index 1, and this is impossible (see 1.4).

3.3 Review of quaternions

We now refresh our memory about quaternions. The *quaternions*

$$x = a_0 + a_1 i + a_2 j + a_3 k ,$$

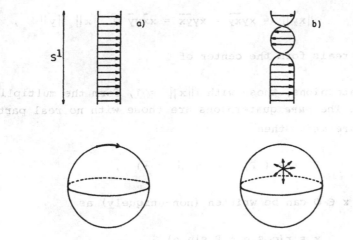

Fig. 2. Graphs of two loops $S^1 \to SO(3)$ and their projections onto $ST(S^2)$

with the product determined by

$$i^2 = j^2 = k^2 = ijk = -1$$

and a_0, ..., a_3 real, are the elements of a non-commutative field $\mathbb{Q}$ (Fig. 3).

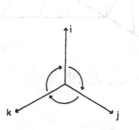

Fig. 3. Law of products

The *real part* of x is $e(x) = a_0$ and the *imaginary part* $v(x)$ *is* $a_1 i + a_2 j + a_3 k$. *The conjugate* $\bar{x}$ of x is $e(x) - v(x)$. We have $\overline{xy} = \bar{y}\bar{x}$. The norm of x is

$$\|x\| = a_0^2 + a_1^2 + a_2^2 + a_3^2$$

and

$$x\bar{x} = \|x\| ,$$

and also

$$\|xy\| = xy\overline{xy} = xy\overline{y}\,\overline{x} = x\overline{x}y\overline{y} = \|x\|\ \|y\|\quad,$$

because the reals form the center of $\mathbb{Q}$.

The *unit* quaternions, those with $\|x\| = 1$, form the multiplicative subgroup S^3. The *pure* quaternions are those with no real part. If P is a *pure unit* then

$$P^2 = P\cdot(-\overline{P}) = -\ \|P\| = -1\quad,$$

hence every $x \in \mathbb{Q}$ can be written (non-uniquely) as

$$x = r(\cos\alpha + P\sin\alpha)\ ,$$

where r is real and P is a pure unit. Symbolically we write $x = re^{\alpha P}$ (Fig. 4).

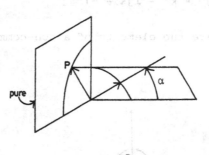

Fig. 4. $e^{\alpha P}$

3.4 Right-helix turns

As for complex numbers, the transformation $x \mapsto e^{\beta Q}x$, where Q is a pure unit, is an orthogonal transformation of $\mathbb{R}^4$.

Exercise 1. Find the equations of $x \mapsto e^{\beta Q}x$ and check that this is an orthogonal transformation which in the plane containing 1 and Q is a rotation of angle β, and in the completely orthogonal plane is a rotation of angle β.

<u>Hint</u>.- Ckeck linearity. Find the images of 1 and Q. We will describe this transformation by its effect on S^3 (note that the norm is preserved).

Represent S^3 by stereographic projection as $\mathbb{R}^3 + \infty$. The 2-sphere of radius 1 ("the unit sphere") is then the set of pure units (Fig. 5 and compare with Fig. 1.16).

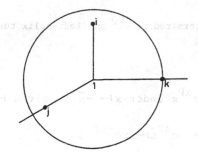

Fig. 5. S^3 in stereographic projection

The action of S^1 on S^3 given by $x \mapsto e^{\beta i}x$, $0 \leq \beta \leq 2\pi$ was studied in 1.5 and we saw that the orbits define the Hopf fibration. Thus $x \mapsto e^{\beta i}x$ is a *right-helix turn* which turns the *equator* jk to the right by β radians, and at the same time pushes the *axis* i upward by β radians. We stress here that neither the equator nor the axis have intrinsic meaning: they depend on the stereographic projection. Since there is a rotation of $\mathbb{R}^4$, fixing 1 and sending i to Q, it is fairly clear that $x \mapsto e^{\beta Q}x$ is a right-helix turn with axis Q which turns to the right and pushes forward by β radians. This movement embeds in an S^1-action whose orbits are a "Hopf fibration" with axis Q (Fig. 6).

3.5 Left-helix turns

We compare $x \mapsto xe^{-\alpha P}$ with $x \mapsto e^{\alpha P}x$ by means of the reflection of $\mathbb{R}^4$ through the pure hyperplane $a_0 = 0$, given by $x \mapsto -\bar{x}$ and we have that

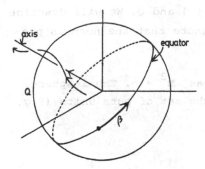
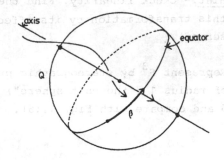

Fig. 6. Right-helix turn determined by $e^{\beta Q}$ and left-helix turn determined by $e^{-\beta Q}$

the conjugation of $x \mapsto e^{\alpha P} x$ under $x \mapsto -\bar{x}$ is given by

$$x \mapsto -\bar{x} \mapsto -e^{\alpha P} \bar{x} \mapsto x e^{-\alpha P} .$$

Now, in the stereographic model of S^3, the reflection $x \to -\bar{x}$ corresponds to the inversion with center 1 and with mirror the unit sphere. Thus $x \mapsto x e^{-\alpha P}$ is a *left-helix turn*, i.e. it turns the equator to the right and pushes α radians backwards along the axis P (Figs. 6, 7). This map embeds in an S^1-action whose orbits are a *left* Hopf fibration.

Exercise 1. Right-helix turns commute with left-helix turns.

3.6 The universal cover of SO(4)

Let P and Q be pure quaternions and α, β be real numbers with $0 \le (\alpha, \beta) \le 2\pi$. Then *the map*

$$\lambda: S^3 \times S^3 \to SO(4)$$

$$(e^{\alpha P}, e^{\beta Q}) \mapsto (x \mapsto e^{\alpha P} x e^{-\beta Q})$$

is a covering with 2 sheets and an epimorphism with kernel $\{(1, 1), (-1, -1)\}$.

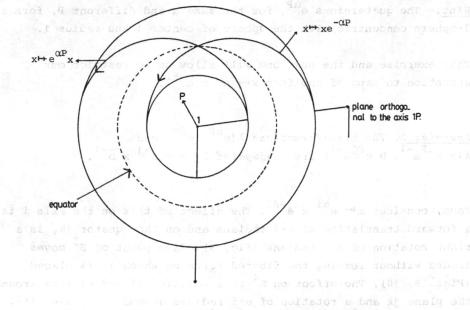

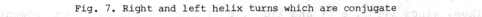

Fig. 7. Right and left helix turns which are conjugate

We first must understand the effect of $x \mapsto e^{\alpha P} x \, e^{-\alpha P}$ on S^3. This fixes the axis 1P pointwise and rotates the equator to the right by 2α radians. Thus, *the map* $x \mapsto e^{\alpha P} x \, e^{-\alpha P}$ *is a rotation of* $\mathbb{R}^3 + \infty$ *around the axis* 1P, *to the right, and of angle* 2α *(Fig. 8)*.

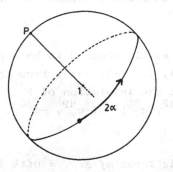

Fig. 8. Rotation of $\mathbb{R}^3 + \infty$, $x \mapsto e^{\alpha P} x \, e^{-\alpha P}$

Exercise 1. Use this to show that given $e^{\alpha i}$ *and* $e^{\alpha P}$ *there exists a unit quaternion* b *such that* $e^{\alpha i} = b \, e^{\alpha P} b^{-1}$.

Hint.- The quaternions $e^{\alpha P}$, for the same α and different P, form a
2-sphere concentric with the sphere of center 1 and radius 1.

This exercise and the next one will allow us to restrict our
attention to maps of the form $x \mapsto e^{\alpha i} x\, e^{-\beta i}$.

Exercise 2. *The transformations* $\lambda(e^{\alpha P}, e^{\beta Q})$ *and*
$\lambda(a\, e^{\alpha P} a^{-1}, b\, e^{\beta Q} b^{-1})$ *are conjugated by* $x \mapsto a^{-1} x\, \bar{b}^{-1}$.

Thus, consider $x \mapsto e^{\alpha i} x\, e^{-\beta i}$. The effect of this on the axis i is
a forward translation of $\alpha - \beta$ radians and on the equator jk, is a
right rotation of $\alpha + \beta$ radians (Fig. 9). Each point of S^3 moves
around without leaving the fibered torus on which it is placed
(Figs. 9, 10). The effect on $\mathbb{R}^4$ is a rotation of $\alpha - \beta$ radians around
the plane jk and a rotation of $\alpha + \beta$ radians around the plane {1i}.
Thus, since $\lambda(e^{\alpha i}, e^{\beta i})$ and $\lambda(e^{\alpha P}, e^{\beta Q})$ are conjugated by an element
of SO(4) we conclude that $\lambda(e^{\alpha P}, e^{\beta Q})$ *is a combination of two
rotations of angles* $\alpha + \beta$ *and* $\alpha - \beta$ *around two orthogonal planes*. Hence
$\lambda(e^{\alpha P}, e^{\alpha Q})$ *is a rotation of angle* 2α *around a plane*. The invariant
plane of $\lambda(e^{\alpha P}, e^{\alpha Q})$ is the one containing 1-PQ and P-Q (check
this!). It follows that every rotation of $\mathbb{R}^4$ around a plane is of
the form $\lambda(e^{\alpha P}, e^{\alpha Q})$, for some α, P and Q. It is well known that
every element of SO(4) is a rotation or the product of two rotations.
Therefore λ is onto and thus every element of SO(4) is of the form
$\lambda(e^{\alpha P}, e^{\beta Q})$.

On the other hand, λ is easily checked to be a differentiable homomorphism with kernel $\{(1, 1), (-1, -1)\}$. From this it follows that
λ is a covering and that the involution of $S^3 \times S^3$ defining the
covering is given by $(e^{\alpha P}, e^{\beta Q}) \mapsto (-e^{\alpha P}, -e^{\beta Q})$.

Exercise 3. *Show that the torus of Fig. 9 with its induced Riemannian
structure has constant curvature (zero)*.

As a corollary of these results we have that *the transformation*

$$\lambda|_\Delta : S^3 \to SO(3) ,$$

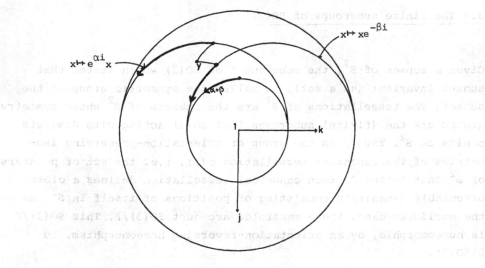

Fig. 9. The combined effect of two helix-turns

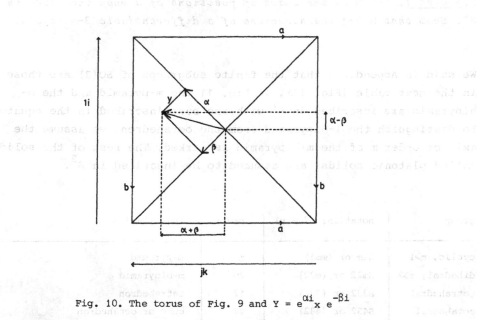

Fig. 10. The torus of Fig. 9 and $Y = e^{\alpha i} x e^{-\beta i}$

given by

$$e^{\alpha P} \mapsto \lambda(e^{\alpha P}, e^{\alpha P})|0 \times \mathbb{R}^3 \quad ,$$

is the universal covering of SO(3). This map is also a homomorphism
with kernel ± 1.

3.7 The finite subgroups of SO(3)

Given a subset of S^2, the subgroup Γ of SO(3) which leaves that subset invariant (as a set), is called the symmetric group of the subset. The tessellations of S^2 are the subsets of S^2 whose symmetry groups are the (finite) subgroups Γ of SO(3) acting with discrete orbits on S^2. Thus Γ is the group of orientation-preserving iso-metries of the canonical tessellation $c(\Gamma)$, i.e. the set of pointers of S^2 that define Γ. Each canonical tessellation defines a closed orientable 3-manifold consisting of positions of itself in S^2. As in the euclidean case, these manifolds are just SO(3)/Γ. This SO(3)/Γ is homeomorphic, by an orientation-reversing homeomorphism, to $\Gamma\backslash$SO(3).

Exercise 1. *Let* M *be the space of positions of a cube inscribed in* S^2. *Show that* M *has the structure of a differentiable 3-manifold.*

We said in Appendix A that the finite subgroups of SO(3) are those in the next table (Fig. 11). In Fig. 11 the m-pyramid and the m-bipyramid are inscribed in S^2 with the base inscribed in the equator. To distinguish the 4-bipyramid from the octahedron, we assume the axis of order m of the m-bipyramid is marked. The rest of the solids, called platonic solids, are assumed to be inscribed in S^2.

group	notation: (ℓ m n)	order	tessellation
cyclic, m$\geqslant$1	Smm or (mm1)	m	m-pyramid
dihedral, m$\geqslant$2	Sm22 or (m22)	2m	m-bipyramid
tetrahedral	S332 or (332)	12	tetrahedron
octahedral	S432 or (432)	24	cube or octahedron
icosahedral	S532 or (532)	60	dodecahedron or icosahedron

Exercise 2. *Check the orders of the groups.*

Exercise 3. *Prove that the above groups have the presentation*

$$|R, S, T : R^n = S^m = T^\ell = RST = 1| \quad ,$$

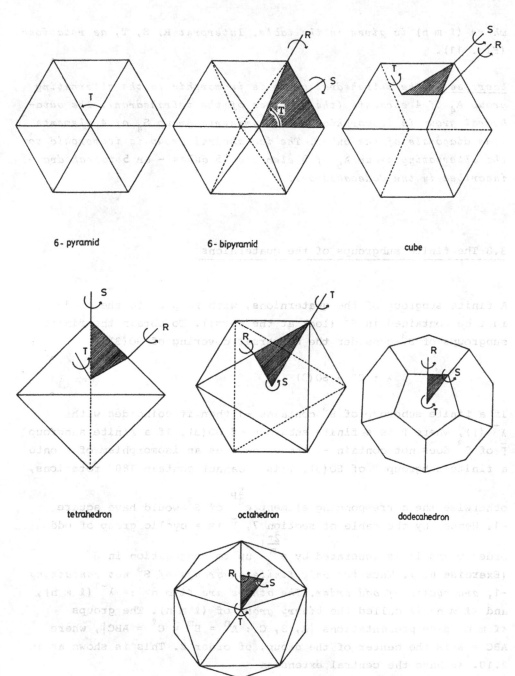

6- pyramid   6- bipyramid   cube

tetrahedron   octahedron   dodecahedron

icosahedron

Fig. 11. The finite subgroups of SO(3)

where (ℓ m n) is given in the table. Interpret R, S, T, as rotations (Fig. 11).

Exercise 4. The tetrahedral group is isomorphic to the alternating group A_4 of 4 elements (the vertices of the tetrahedron). The octahedral group is isomorphic to the symmetric group S_4 of 4 elements (the diagonals of the cube). The icosahedral group is isomorphic to the alternating group A_5 of 5 elements (5 cubes - or 5 tetrahedra - inscribed in the dodecahedron).

3.8 The finite subgroups of the quaternions

A finite subgroup of the quaternions, with respect to the product, must be contained in S^3 (look at the norm!). To obtain the finite subgroups of S^3 consider the universal covering of SO(3)

$$\lambda : S^3 \to SO(3) .$$

If a finite subgroup of S^3 contains -1 then it coincides with $\lambda^{-1}(\Gamma)$, where Γ is a finite subgroup of SO(3). If a finite subgroup $\tilde{\Gamma}$ of S^3 does not contain -1 then λ defines an isomorphism of $\tilde{\Gamma}$ onto a finite subgroup Γ of SO(3). This Γ cannot contain 180° rotations, otherwise the corresponding element $e^{\frac{\pi}{2}P}$ of S^3 would have square -1. Hence, by the table of section 7, $\tilde{\Gamma}$ is a cyclic group of odd order m and it is generated by $e^{\frac{2\pi}{m}i}$ up to conjugation in S^3 (Exercise 6.1). Thus *the only finite subgroups of S^3 not containing -1, are cyclic of odd order. The others are* <ℓ m n> := λ^{-1}(ℓ m n), and <ℓ m n> is called the *binary group* of (ℓ m n). The groups <ℓ m n> have presentations $|A, B, C : A^n = B^m = C^\ell = ABC|$, where ABC = Z is the center of the group, of order 2. This is shown as in 2.10. We have the central extension

$$1 \to \mathbf{Z}_2 \to \text{<ℓ m n>} \overset{\lambda}{\to} (ℓ \; m \; n) \to 1$$

and defining the function φ:(ℓ m n) → <ℓ m n> by

$$(x \mapsto e^{\alpha P} x e^{-\alpha P}) \mapsto e^{\beta P},$$

where $0 \leq \beta < \pi$ and $\alpha \equiv \beta \mod \pi$, the elements $\varphi(R) := A$, $\varphi(S) := B$ and $\varphi(T) := C$ generate the group. The elements A, B, C are $e^{\frac{\pi}{n}R}$, $e^{\frac{\pi}{m}Q}$, $e^{\frac{\pi}{\ell}P}$, where R, Q, P are the vertices of the spherical triangle of Fig. 12 contained in the 2-sphere of pure quaternions. In fact $\lambda(e^{\frac{\pi}{\ell}P})$, for instance, equals $x \mapsto e^{\frac{\pi}{\ell}P} x e^{-\frac{\pi}{\ell}P}$ which is a rotation of angle $2\pi/\ell$ around the axis $1P$ (compare Fig. 11).

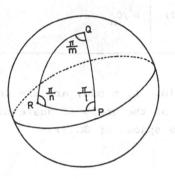

Fig. 12. The spherical triangle $(\ell\ m\ n)$

Thus we have the following table:

group	$<\ell\ m\ n>$	order
cyclic	C_n and $<m\ m\ 1>$	n odd and 2m
binary dihedral	$<m\ 2\ 2>$	4m
binary tetrahedral	$<332>$	24
binary octahedral	$<432>$	48
binary icosahedral	$<532>$	120

The finite subgroups of quaternions act without fixed points on S^3 because $e^{\alpha P} x = x$ implies $e^{\alpha P} = 1$. Thus the manifolds of tessellations in the spherical case are the following:

Manifold	Notation	Definition	π_1	Tessellation
Lens space	L(2m, 1)	$S^3/\langle mm1\rangle = SO(3)/(mm1)$	C_{2m}	m-pyramids
Prism space	[m, 1]	$S^3/\langle m22\rangle = SO(3)/(m22)$	$\langle m22\rangle$	m-bipyramids
Octahedral space	M(3 3 2)	$S^3/\langle 332\rangle = SO(3)/(332)$	$\langle 332\rangle$	tetrahedra
	M(4 3 2)	$S^3/\langle 432\rangle = SO(3)/(432)$	$\langle 432\rangle$	cubes or octahedra
Dodecahedral space	M(5 3 2)	$S^3/\langle 532\rangle = SO(3)/(532)$	$\langle 532\rangle$	dodecahedra or icosahedra
Lens space	L(m, 1) m odd	S^3/C_m	C_m	is not a manifold of tessellations

Thus the lens spaces L(m, 1), m odd, are the only manifolds which are homogeneous spaces of the group of quaternions of norm 1, S^3, but are not homogeneous spaces of SO(3).

Exercise 1. Compare the following three groups of order 120: $C_2 \times (532)$, S_5 and $\langle 532\rangle$.

Exercise 2. Visualize the elements of the fundamental groups of manifolds of tessellations using paths of rotations which start at the identity map and end fixing some base tessellation (compare 2.9).

Exercise 3. Show that $\langle 532\rangle$ is a perfect group (i.e. one which equals its commutator subgroup). Thus M(532) is a homology 3-sphere (the first encountered!) found by Poincaré (and which is named after him).

Hint.- If the fundamental group of a closed, oriented 3-manifold M is perfect, then $H_1(M; \mathbb{Z}) = 0$, and then also (by the Poincaré duality) $H_2(M; \mathbb{Z}) = 0$. From this it follows that M is homologically S^3 (i.e. a *homology 3-sphere*).

3.9 Description of the manifolds of tessellations

We now describe the manifolds $S^3/\tilde{\Gamma}$, or rather their orientation-reversing homeomorphs $\tilde{\Gamma}\backslash S^3$, by suitable polyhedra with pairs of faces identified. We start with $C_m\backslash S^3$, where C_m is the set of quaternions $e^{k\frac{2\pi}{m}i}$, $k = 0, 1, \ldots, m-1$ of S^3. They lie on the 1i axis and, via λ, they define a cyclic group of rotations of the unit sphere S^2, generated by the rotation of axis 1i and angle $\frac{4\pi}{m}$ (Fig. 13). The distribution of the quaternions $e^{k\frac{2\pi}{m}i}$ on axis 1i is clearly determined by the stereographic projection (Fig. 14).

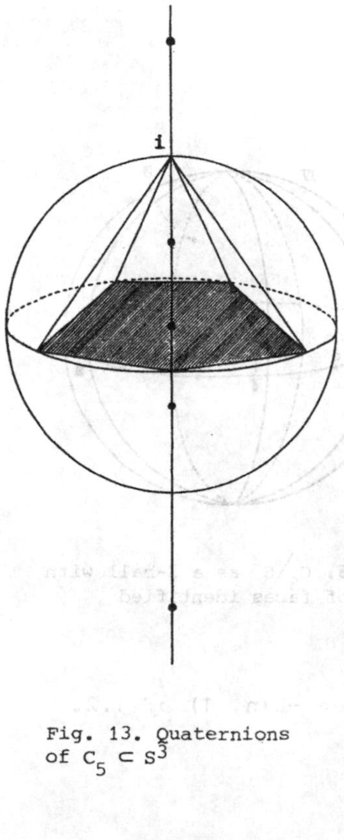

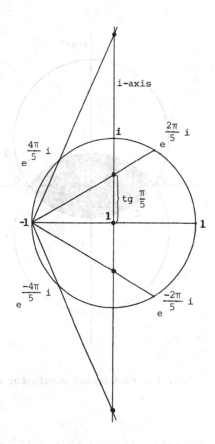

Fig. 13. Quaternions
of $C_5 \subset S^3$

Fig. 14. Stereographic projection of
the 1-sphere 1i from the south pole
-1

The group C_m acts on the left on S^3 by $x \mapsto e^{k\frac{2\pi}{m}i} x$. This is a right-helix turn with axis $1i$ and angle $\frac{2\pi}{m}k$. To obtain a fundamental domain for this action, and subsequently the quotient manifold $C_m \backslash S^3$ of right classes $\{C_m x\}$, we take a 3-dimensional half-space E of $\mathbb{R}^4$ whose boundary is a 2-space C which is invariant under C_m. Thus C cuts S^3 in a fiber of the corresponding Hopf fibration associated to the right-helix turn, and E cuts in a 2-dimensional half-sphere d (Fig. 15). The m images of d divide S^3 into lenses of axis $1i$. Each lens is a fundamental domain for the action. In Fig. 15 we have taken d and one of its images so that together they are symmetric with respect to the disk D^2 passing through 1. Then the manifold $C_m \backslash S^3$ is the result of identifying the bottom and top of that lens after having performed a rotation of $\frac{2\pi}{m}$ (Fig. 16).

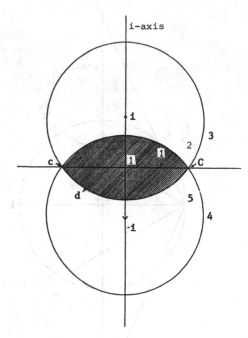

Fig. 15. Fundamental domain for C_5

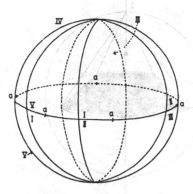

Fig. 16. $C_5 \backslash S^3$ as a 3-ball with pairs of faces identified

Exercise 1. Show that $C_m \backslash S^3$ is the lens space $-L(n, 1)$ of 1.2.

<u>Hint.</u>- Decompose $C_m \backslash S^3$ into two solid tori which are neighbourhoods of the axis 1i and of the 1-sphere jk.

3.10 Prism manifolds

Now we describe the manifold $<m222> \backslash S^3$. As before we first describe the quaternions which constitute the group $<m222>$. Recall that $<m22>$ is the binary group of the group of orientation-preserving isometries of S^2 fixing the m-bipyramid.

We consider first the particular case m=2. The 2-bipyramid has 3 axes of symmetry E_1, E_2, E_3 and the 8 quaternions of $<222>$ lie on them. Since the rotations around the axes E_i are 180° rotations, it follows that the quaternions of $<222>$ are precisely $\{\pm1, \pm i, \pm j, \pm k\}$ (Fig. 17). This is the so-called *quaternionic group* presented by

$$|i, j, k : i^2 = j^2 = k^2 = ijk| \quad .$$

The 8 quaternions are the vertices of a solid, analogous to the octahedron, which we call a *hyperoctahedron*. This solid can be thought of as a 4-dimensional solid ("polytope") inscribed in $S^3 \subset \mathbb{R}^4$ or as a "curved" polytope obtained by projecting the recti-linear polytope onto S^3 from the origin of $\mathbb{R}^4$. This solid is a tessellation of S^3, in exactly the same way as the set of m lenses of the last section forms a tessellation of S^3.

Given a vertex of the hyperoctahedron, a point of S^3, whose distance (with respect to the metric of S^3) to that vertex is less than or equal to its distance to the other vertices, is said to lie in the *Dirichlet region* of that vertex. This region is a (curved) cube, and the set of these 8 cubes forms a polytope which is analogous to the cube and is called *hypercube* or *tesseract* (Fig. 18).

Thus the cubes of the hypercube serve as fundamental domains for the action of the quaternionic group on S^3. Then the manifold $<222> \backslash S^3$, called *quaternionic space* (we will see that it is (On1|2); compare Exercise 1.7.2), is the result of identifying opposite faces of a cube under a right-helix turn of angle $\pi/2$ (Fig. 19).

116

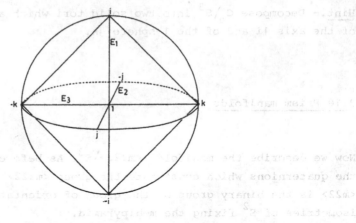

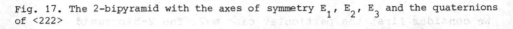

Fig. 17. The 2-bipyramid with the axes of symmetry E_1, E_2, E_3 and the quaternions of <222>

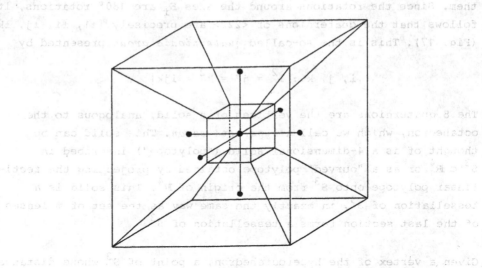

Fig. 18. The hypercube and the vertices of the hyperoctahedron

The general case <m22> is similar. The subgroup $C_m \triangleleft (m22)$, which corresponds to the inclusion of a m-pyramid coaxial with the m-bipyramid, gives rise to the subgroup $C_{2m} \triangleleft <m22>$. The 2m quaternions C_{2m} are placed on the principal axis of the pyramid. The remaining 2m quaternions lie on the equator of S^2, equally spaced, and on the axis of 2-fold symmetry of the m-bipyramid (Fig. 20).

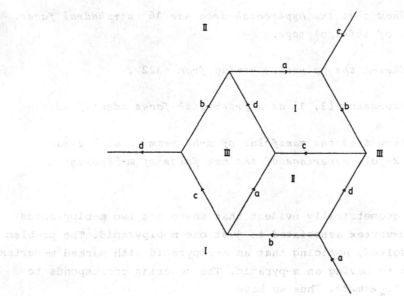

Fig. 19. The quaternionic space as a cube (under the book) with opposite faces identified

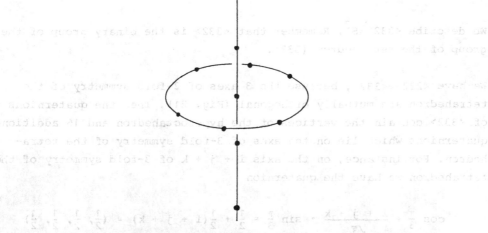

Fig. 20. The quaternions of <322>

Then the *prism space* $[m, 1]$ = $-$<m22>$\backslash S^3$ is the result of identifying opposite faces of a 2m-prism: base and top, under a right-helix turn of angle π/m, and side faces under a right-helix turn of angle $\pi/2$.

Exercise 1. Show that the hyperoctahedron has 16 tetrahedral faces. Draw a sketch of this polytope.

Exercise 2. Sketch the polytopes coming from <322>.

Exercise 3. Represent [3, 1] as a prism with faces identified.

Exercise 4. Show that the manifolds of m-bipyramids with a marked m-vertex are 2-fold coverings of the manifolds of m-bipyramids.

Hint.- It is geometrically evident that there are two m-bipyramids with marked m-vertex associated to just one m-bipyramid. The problem can also be solved, noticing that an m-bipyramid with marked m-vertex is equivalent to having an m-pyramid. The covering corresponds to the subgroup $C_{2m} \triangleleft <m22>$. Thus we have

$$L(2m, 1) \xrightarrow{2:1} [m, 1] \quad .$$

3.11 The octahedral space

We describe $<332>\backslash S^3$. Remember that <332> is the binary group of the group of the tetrahedron (332).

We have $<222> \triangleleft <332>$, because the 3 axes of 2-fold symmetry of the tetrahedron are mutually orthogonal (Fig. 21). Then the quaternions of <332> contain the vertices of the hyperoctahedron and 16 additional quaternions which lie on the axes of 3-fold symmetry of the tetrahedron. For instance, on the axis i + j + k of 3-fold symmetry of the tetrahedron we have the quaternion

$$\cos \frac{\pi}{3} + \frac{i + j + k}{\sqrt{3}} \cdot \sin \frac{\pi}{3} = \frac{1}{2} + \frac{1}{2}(i + j + k) = (\tfrac{1}{2},\ \tfrac{1}{2},\ \tfrac{1}{2},\ \tfrac{1}{2})$$

(Fig. 22). Thus we readily see that these 16 quaternions are just the vertices of the hypercube. Hence <332> is the union of the vertices of the hyperoctahedron and the vertices of the hypercube. This union forms the vertices of a polytope called a *24-cell*, which clearly has 24 faces, which are octahedra.

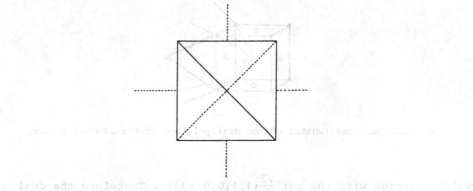

Fig. 21. The axes of 2-fold symmetry of the tetrahedron

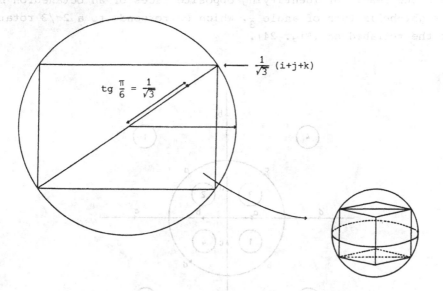

$$\operatorname{tg} \frac{\pi}{6} = \frac{1}{\sqrt{3}}$$

$$\frac{1}{\sqrt{3}}\,(i+j+k)$$

Fig. 22. Quaternions which are placed on a 3-fold axis of the tetrahedron

The fundamental domain for the action of <332> on S^3 is just a cell of the dual polytope of the 24-cell, having its vertices in the middle of the faces of the 24-cell. These vertices are the following 24 quaternions (see Fig. 23):

$$\{\frac{\sqrt{2}}{2}(\pm 1,\ \pm 1,\ 0,\ 0),\ \frac{\sqrt{2}}{2}(0,\ 0,\ \pm 1,\ \pm 1),\ \frac{\sqrt{2}}{2}(0,\ \pm 1,\ \pm 1,\ 0),$$

$$\frac{\sqrt{2}}{2}(\pm 1,\ 0,\ \pm 1,\ 0),\ \frac{\sqrt{2}}{2}(0,\ \pm 1,\ 0,\ \pm 1),\ \frac{\sqrt{2}}{2}(\pm 1,\ 0,\ 0,\ \pm 1)\}$$

120

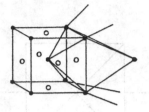

Fig. 23. Vertices (white) of the dual polytope of the 24-cell (black)

which coincide with the set $\frac{\sqrt{2}}{2}(1,1,0,0)<332>$. Therefore the dual of the 24-cell is the 24-cell. Thus the fundamental domain for the action of <332> is an octahedron. The *octahedral space* $<332>\backslash s^3$ is then the result of identifying opposite faces of an octahedron by a right-helix turn of angle $\frac{\pi}{3}$, which corresponds to a $2\pi/3$ rotation of the tetrahedron (Fig. 24).

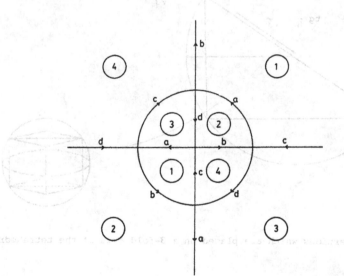

Fig. 24. The octahedral space

Exercise 1. Show that the set of vertices of the hypercube is the union of the set of vertices of two hyperoctahedra.

Hint.- <222>◄<332> has index 3. As representatives of the other two classes take, for instance, $v = (\frac{1}{2}, \frac{1}{2}, \frac{1}{2}, \frac{1}{2})$ and v^2. Thus the hypercube is the union of the classes v<222> and v^2<222>.

Exercise 2. From the polyhedra with identified faces determine the homology with integer coefficients of the prism spaces and of the octahedral space. Draw generators for the homology and check Poincaré duality geometrically.

Exercise 3. Show that the quaternionic space is a regular 3-fold covering of the octahedral space. Visualize it in terms of tessellations.

<u>Hint</u>.- The covering corresponds to <222>◀<332>. The tessellation formed from a tetrahedron with two opposite edges equally marked has the same symmetry as the 2-bipyramid. The space of these tessellations covers the space of tetrahedral tessellations 3 to 1 because there are 3 pairs of opposite edges in a tetrahedron.

Exercise 4. The space of tessellations formed from tetrahedra with a marked vertex is a 4-fold covering of the octahedral space. Is it a regular covering? Determine the associated subgroup and the covering.

<u>Hint</u>.- The cover is L(6, 1) and is not regular: lift a path of tessellations giving 1/3 of a turn around a ternary axis.

3.12 The truncated-cube space

We now study <432>\$S^3$ where (432) is the group of the cube. Note that since it is possible to inscribe a tetrahedron in a cube we have that (332) is a subgroup of index 2 of (432). Hence <332>◀<432>. Therefore, <432>, having order 48, coincides with the vertices of the 24-cell together with the vertices of another 24-cell v<332>, where v is, for instance, a quaternion representing a π/2 rotation around a 4-fold axis of the cube. This v might be $\frac{\sqrt{2}}{2}(1,1,0,0) = e^{\frac{\pi}{4}i}$, for instance. Hence <432> is the union of the vertices of the 24-cell with the vertices of the dual 24-cell (Fig. 25).

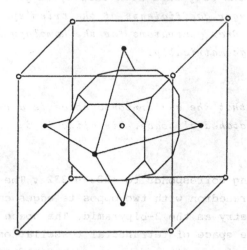

Fig. 25. Vertices of <432> and truncated cube of the dual polytope

The dual polytope, whose cells serve as fundamental domains for the action of <432> on S^3, is formed from *truncated cubes*. In fact the *truncated cube space* <432>\S^3 is obtained identifying opposite faces of this truncated cube: *octagonal* faces through right-helix turns of angle $\frac{\pi}{4}$, and triangular faces by right-helix turns of angle $\frac{\pi}{3}$ (Fig. 26).

Exercise 1. Find the homology groups of this manifold using Fig. 26.

Exercise 2. The octahedral space is a 2-fold covering of the truncated-cube space. Visualize this in terms of tessellations.

Exercise 3. The manifold of cubes with a diagonal marked is a 4-fold covering of the truncated-cube space. Is it regular? Determine the associated subgroup and the covering space.

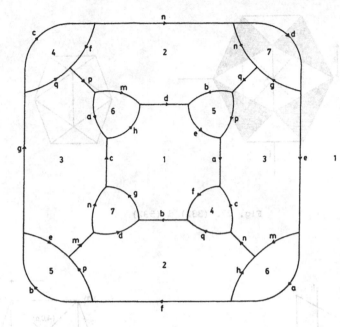

Fig. 26. Truncated-cube space

Hint.- The group of symmetry of a cube with a marked diagonal coincides with the group of the 3-bipyramid. The covering is [3, 1] and is not regular.

3.13 The dodecahedral space

Consider the group (532) of symmetries of the dodecahedron. We want to describe the manifold $<532>\backslash s^3$.

First notice that $<332>$ is a subgroup of $<532>$ because it is possible to inscribe five cubes in a dodecahedron. The subgroup is not normal because $(532) \cong A_5$ is simple (Fig. 27).

We want to obtain the quaternions of $<532> - <332>$. To do this, remember that the vertices of three cardboard golden-rectangles (i.e. the ratio of sides is the golden ratio $\frac{1 + \sqrt{5}}{2}$), symmetrically inter-crossed, form the vertices of the icosahedron (Fig. 28).

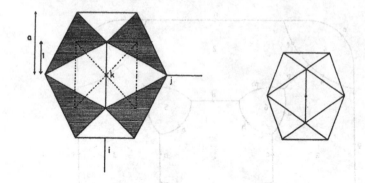

Fig. 27. (332) ≤ (532)

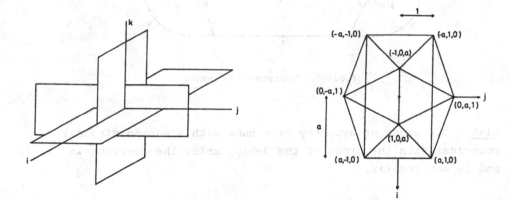

Fig. 28. The vertices of the icosahedron, $a = \dfrac{1 + \sqrt{5}}{2}$

Thus the pure quaternions defining the pentagonal axis are $\dfrac{1}{\sqrt{1 + a^2}}(1,0,a)$, etc... Therefore the quaternion defining the $\dfrac{2\pi}{5}$ rotation around the axis $(1, 0, a)$ is

$$(1) \quad \cos\frac{\pi}{5} + \sin\frac{\pi}{5}\left(\frac{1}{\sqrt{1 + a^2}}(1, 0, a)\right) = \frac{a}{2} + \frac{\sqrt{1 + a^{-2}}}{2}\left(\frac{1}{\sqrt{1 + a^2}}(1, 0, a)\right) =$$

$$= \frac{a}{2} + \frac{a^{-1}}{2}(1, 0, a) = \frac{1}{2}(a, a^{-1}, 0, 1).$$

In this way it is easy to obtain the quaternions of <532> - <332>, and we obtain the family $\frac{1}{2}((\pm a, \pm 1, \pm a^{-1}, 0))$, where the double parentheses indicates that we have to take all the even permutations of the components.

The vertices of <532> whose distance to $(1,0,0,0)$ is the smallest possible are those with first coordinate equal to $\frac{a}{2}$, as is easily checked. These vertices are $\frac{1}{2}(a, 0, \pm 1, a^{-1})$, $\frac{1}{2}(a, \pm a^{-1}, 0, \pm 1)$, $\frac{1}{2}(a, \pm 1, \pm a^{-1}, 0)$, and they have the form (1), hence they project stereographically onto the vertices of an icosahedron homothetic to the one of Fig. 28. This icosahedron is divided into 20 tetrahedra obtained by joining $(1,0,0,0)$ with its vertices. These tetrahedra are regular (in $\mathbb{R}^4$) and thus the vertices of <532> form a polytope with 600 tetrahedral faces (the *600-cell*). The dual polytope has 120 cells (the 120-cell) which are dodecahedra, which serve as fundamental domains for the action of <532> on S^3. Thus the *dodecahedral space* <532>\$S^3$ is the result of identifying opposite faces of a regular dodecahedron by a right-helix turn of angle $\frac{\pi}{5}$ (Fig. 29).

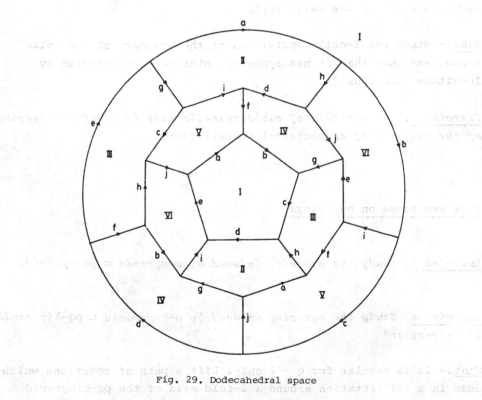

Fig. 29. Dodecahedral space

Exercise 1. Show that <332> decomposes <532> into 5 left classes; namely x^i<332> i=0, 1, 2, 3, 4, where x is of order 5 (for instance $x = \frac{1}{2}(a^{-1}, a, 1, 0)$).

Exercise 2. Show that the polytope formed by the vertices of <532> consists of 600 regular tetrahedra.

Exercise 3. Use Fig. 29 to show that <532>\S^3 is a homology sphere ("Poincaré sphere").

Exercise 4. Draw layers of dodecahedra around a fixed dodecahedron of the 120-cell and check that, when 60 of them are used, the resulting 3-ball has its boundary decomposed into 2-cells in such a way that it has antipodal symmetry ("decomposition of SO(3)").

Exercise 5. Show that the 120-cell is the union of two solid tori, each made of 60 dodecahedra, as follows. Take a pile of 10 dodecahedra making a ring. The boundary of the ring has 10 circular notches, and in each one of them fit 5 dodecahedra. These 60 dodecahedra form one of the solid tori.

Hint.- Study the 2-cell subdivision of the boundary of the solid torus, and show that it has symmetry interchanging meridian by longitude (see [Cox1]).

Exercise 6. The manifold of cubic tessellations is a 5-fold covering of the manifold of dodecahedral tessellations.

3.14 Exercises on coverings

Exercise 1. Study the covering induced by p-pyramid ⊂ pq-pyramid.

Exercise 2. Study the covering induced by p-bipyramid ⊂ pq-bipyramid. Is it regular?

Hint.- It is regular for q = 2 only. Lift a path of rotations which ends in a 180° rotation around a 2-fold axis of the pq-bipyramid (Fig. 30).

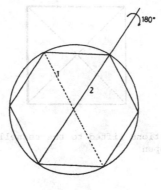

Fig. 30. Lifting the 180° rotation to the 2-pyramid of base 1, the path is not closed, but it is closed if it is lifted to 2

Exercise 3. Show that the group of covering transformations of the covering induced by p-pyramid ⊂ pq-bipyramid is the dihedral group (q22).

Hint.- $\langle pq, 2, 2\rangle = |A, B : A^{pq} = B^2 = (AB)^2|$ and A^q generates C_{2p}.

Exercise 4. If p is odd, show that $\langle pq, 2, 2\rangle$ has a normal subgroup C_p such that $\langle pq, 2, 2\rangle/C_p \cong \langle q, 2, 2\rangle$.

Hint.- Since $B^{-1}AB^{-1} = A^{-1}$, A^{2q} generates a normal subgroup.

Exercise 5. There is a 4-fold covering $L(p, 1) \to [p, 1]$ if p is odd.

Exercise 6. Show geometrically that the centers of $\langle 332\rangle$ and $\langle 432\rangle$ are normal subgroups. Determine the corresponding coverings and the groups of covering transformations.

Hint.- Two adjacent edges on a face of the tetrahedron or of the cube form an asymmetric tessellation of S^2. The space of these tessellations is $L(2, 1) \cong SO(3)$ which is a 12-fold cover of the octahedral space and a 24-fold cover of the truncated cube space (Fig. 31). The groups of covering transformations are (332) and (432).

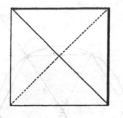

Fig. 31. The loops of rotations lifted to the tessellations of two legs are either all closed or all open

Exercise 7. Study the covering corresponding to the inclusion of the tessellation, consisting of two opposite edges of a cube, in the cube itself. Is it regular? Find the group of covering translations.

Solution.- It corresponds to <222> ◁ <432>. Moreover <432>/<222> ≅ (322).

Exercise 8. From the last exercise it follows that the quaternionic space admits a free action of the symmetric group of three elements. Describe this action in the model of Fig. 19.

Hint.- Figure 32.

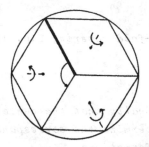

Fig. 32. The group S_3 acts freely on the quaternionic space

Exercise 9. Check the diagram of coverings.

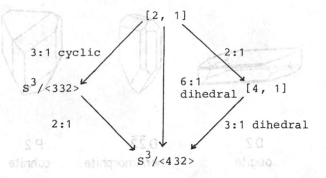

Hint.- The 3-fold covering [4, 1] → $S^3/<432>$ corresponds to the
inclusion of a 4-bipyramid (with a marked diagonal) in an octahedron.
The monodromy group is the dihedral (322).

Exercise 10. *Show that, in contrast to the euclidean case, no*
manifold of spherical tessellations is a bundle over S^1.

Hint.- There is a short exact sequence $1 \to \pi_1(F) \to \pi_1(M) \to \mathbb{Z} \to 0$.

3.15 Involutions on the manifolds of spherical tessellations

Allowing isometries of the 2-sphere which reverse orientation, such
as reflections in great circles, there are a number of new tessellations, which correspond to finite subgroups of O(3), up to conjugation in O(3) (see Appendix A, for the complete list).

The four "crystals" of Fig. 33 have the same "orientation-preserving"
symmetry, namely the symmetry of the 2-pyramid, but they define four
non-conjugate finite subgroups of O(3).

Take a fixed diametral plane E in S^2 and call r the reflection
through E. The manifold of tessellations defined by any of the
crystals of Fig. 33 is the lens space L(4, 1). The reflection r
applied to S22 produces the "enantiomorph" crystal which is different
from S22, but when r is applied to D2, D$\overline{2}\overline{2}$ or P2 we see that the
crystal obtained is the same. Thus r defines 3 involutions on
L(4, 1). The construction of these involutions and the quotient

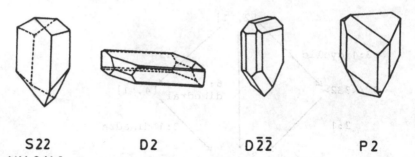

S22	D2	D$\bar{2}\bar{2}$	P2
NH$_3$C$_2$H$_5$I	augite	hemimorphite	cahnite

Fig. 33. C$_2$ is the group of orientation-preserving symmetry

spaces will be done in the next chapter, after introducing Seifert manifolds.

Other examples of the situation are shown in Figs. 34, 35, 36, 37, 38. The manifolds are, respectively, L(2, 1), the octahedral space, the space of truncated-cubes, the prism manifold [3, 1], and the dodecahedral space.

S	D	P
SrH$_2$(C$_4$H$_4$O$_6$)$_2$·4H$_2$O	hilgardite	albite

Fig. 34. Orientation-preserving group: trivial

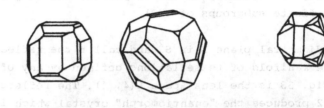

S 332	D $\bar{3}\bar{3}\bar{2}$	D 3$\bar{2}$
Sodium Clorate	Eulytine	Pyrite

Fig. 35. Orientation-preserving group: (332)

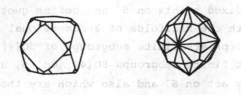

S 432
Cu₂O

D 432
silver

Fig. 36. Orientation-preserving group: (432)

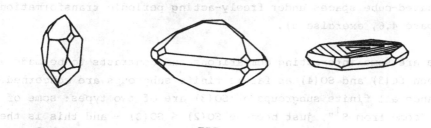

S 322
quartz

D 322
benitoite

D 23
hematite

Fig. 37. Orientation-preserving group: (322)

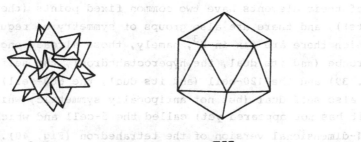

S 532

D 532

Fig. 38. Orientation-preserving group: (532)

3.16 The groups $\tilde{\Gamma}$ as groups of tessellations of S^3

We have already observed that the groups C_m, <m22>, <332>, <432> and <532> act on S^3 fixing some particular 3-dimensional tessellation.

They act without fixed points on S^3 and define quotient manifolds
which coincide with the manifolds of 2-dimensional spherical tessellations. Those groups are finite subgroups of SO(4) and we would
like to know, what finite subgroups SO(4) admits, up to conjugation
in SO(4), how they act on S^3 and also which are the quotient
orbifolds. In particular, are there more 3-manifolds than the
ones already encountered? There are, in fact, more and all of
them are *spherical manifolds*, i.e. they have a Riemannian metric
(inherited from S^3) of constant positive sectional curvature. The
new manifolds are more lens spaces, a generalization of the prism
manifolds and quotients of the octahedral, dodecahedral and
truncated-cube spaces under freely-acting periodic transformations
(compare 4.6, exercise 5).

There are some interesting comparisons and contrasts to be made
between SO(3) and SO(4) as far as finite subgroups are concerned. For
instance all finite subgroups of SO(3) are of two types: some of
them "come from S^1", just because SO(2) $\leq$ SO(3) - and this is the
case of the cyclic groups -, or because O(2) also lies in SO(3)
(by extending orientation-reversing transformations of S^1 to
rotations of SO(3)), giving rise to the dihedral groups; the rest of
the finite subgroups of SO(3) are the symmetric groups of certain
regular polyhedra called platonic solids. In SO(4) there are also
groups "coming from S^2" (suspensions of finite subgroups of SO(3):
all of their elements have two common fixed points (the suspension
points)), and there are also groups of symmetry of regular polytopes,
of which there are six in S^3, namely, those already encountered: the
hypercube (and its dual, the hyperoctahedron), the self-dual 24-cell
(Fig. 39) and the 120-cell (and its dual, the 600-cell); and a new
one, also self-dual (but not antipodally symmetric, which explains
why it has not appeared yet) called the 5-cell and which is just
the 4-dimensional version of the tetrahedron (Fig. 40). But in SO(4)
there are also finite subgroups not of this type, as for instance the
group <432>. Another difference between SO(3) and SO(4) is that the
same abstract group appears as a subgroup of SO(4) in different
(i.e. non-conjugate) ways. This happens for instance with the cyclic
groups, which quotient S^3 to different lens spaces, and also with
the group A_5 = (532), which acts on S^3 in two different ways: one
as the suspension of (532) $\leq$ SO(3), another as the full group of
symmetries of the 5-cell.

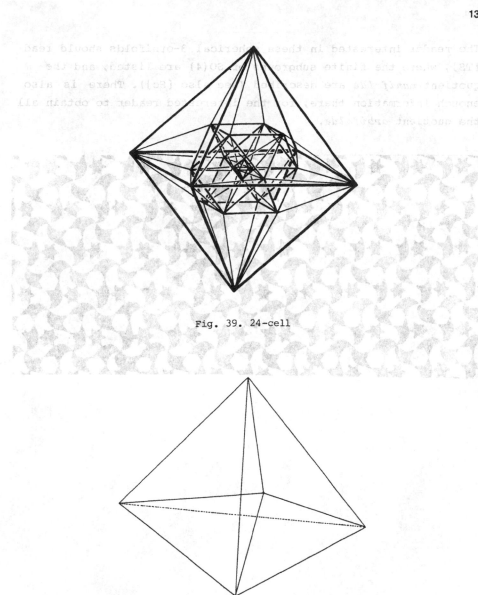

Fig. 39. 24-cell

Fig. 40. The 5-cell

These remarks will make the reader aware that the problem of
classifying the finite subgroups of SO(4) is not at all easy, but,
at the same time, they tend to show that the problem is also very
beautiful.

The reader interested in these spherical 3-orbifolds should read [TS], where the finite subgroups of SO(4) are listed, and the quotient *manifolds* are described (see also [Sc]). There is also enough information there, for the interested reader to obtain all the quotient *orbifolds*.

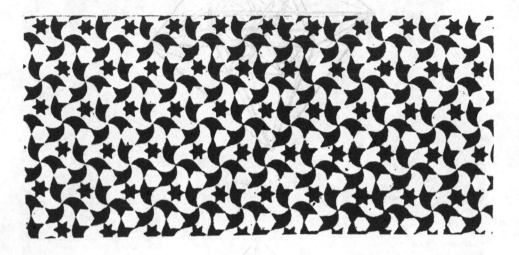

Chapter Four: Seifert Manifolds

"Mire vuestra merced, respondió Sancho, que aquellos
que allí se parecen, no son gigantes, sino molinos
de viento, y lo que en ellos parecen brazos son las
aspas."

*"Take care, your worship," said Sancho; "those
things over there are not giants but windmills, and
what seem to be their arms are the sails, which are
whirled round in the wind and make the millstone
turn."*
Cervantes, Don Quixote, Part I, Ch. VIII, Of the
valorous Don Quixote's success in the dreadful and
never before imagined Adventure of the Windmills,
with other events worthy of happy record.

The spherical bundles of surfaces as well as the manifolds of tessellations are examples of Seifert manifolds. Using the language of
orbifolds, a Seifert manifold is a manifold (i.e. an orbifold with
empty singular set) that fibers over a 2-orbifold whose singular
points form a discrete set (and have cyclic isotropy groups). The
manifolds of tessellations are the spherical bundles of such
2-orbifolds.

In this chapter, we study the orientable Seifert manifolds and we
give a set of invariants that classify them up to orientation-
preserving and fiber-preserving equivalence. We closely follow the
original paper [S], which we recommend as the best source for this
topic.

We devote part of the chapter to a careful proof that the manifolds
of tessellations are Seifert manifolds and we find their invariants.

We close the chapter by discussing involutions on a Seifert manifold. This topic is used to understand the effect, on a manifold of
tessellations, of the involution defined by reflection through an
axis of E^2 or S^2.

4.1 Definition

A closed, orientable, connected 3-manifold is a *Seifert manifold* if it is a union of fibers, all homeomorphic to the 1-sphere S^1, such that each point of the manifold belongs to exactly one fiber, and such that each fiber has a solid torus neighbourhood, made of fibers which are not meridians of it. In fact, a Seifert-fiber structure is equivalent to a foliation by circles. This is a highly non-trivial theorem of D.B.A. Epstein.

It is convenient to introduce the following model of an *oriented fibered solid torus of type* β/α, $0 < \beta < \alpha$ (compare [BS]). Take

$$V(\beta/\alpha) = (D^2 \times [0, 1])/R$$

where R identifies $(x, 1)$ with $(r(\beta/\alpha)x, 0)$, where

$$r(\beta/\alpha): D^2 \to D^2$$

is rotation of angle

$$2\pi\beta*/\alpha, \quad \beta*\beta \equiv 1 \bmod \alpha, \quad \beta* \in (0,\alpha) \quad .$$

The orientation of $V(\beta/\alpha)$ is given by the orientation of D^2 followed by the orientation of $[0, 1]$ (Fig. 1). Thus $V(\beta/\alpha)$ is a mapping torus of D^2 with monodromy of period α. Topologically, $V(\beta/\alpha)$ is a solid torus on which there is an S^1-action whose orbits are made of α vertical segments of $D^2 \times [0, 1]$, all of them different, except for the orbit passing through the center of D^2 where the α segments coincide. Thus $V(\beta/\alpha)$ is fibered over the disk

$$V/S^1 \cong E^2$$

(Fig. 2) with an *exceptional fiber* corresponding to the axis of the solid torus (in the language of orbifolds, $V(\beta/\alpha)$ is a S^1-bundle over the orbifold E^2 in which 0 has isotropy group cyclic of order α). Any other (*general*) fiber has a neighbourhood $U \times S^1$, where U is a neighbourhood in the base E^2 not containing 0 (Fig. 3).

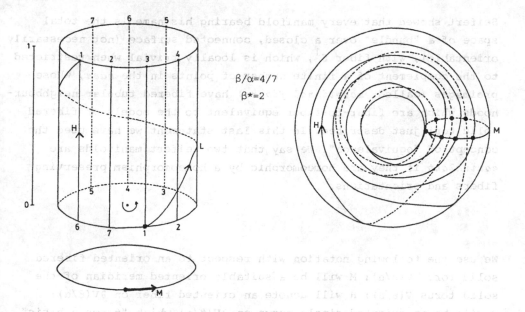

$\beta/\alpha = 4/7$

$\beta* = 2$

Fig. 1. The oriented fibered torus V(4/7)

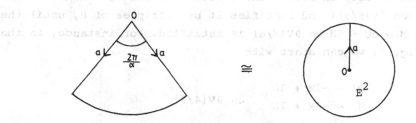

Fig. 2. The base of the fibered solid torus

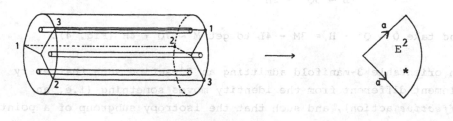

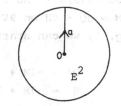

Fig. 3. Tubular neighbourhood of a general fiber

138

Seifert showed that every manifold bearing his name is the total
space of a "bundle" over a closed, connected surface (not necessarily
orientable), with fiber S^1, which is locally trivial when restricted
to the complement of a finite number of points in the *base*, whose
preimages (called *exceptional fibers*) have fibered tubular neighbourhoods which are fiber-to-fiber equivalent to the models of fibered
solid tori just described. In this last statement we have used the
concept of "equivalence". We say that two Seifert manifolds are
equivalent if they are homeomorphic by a homeomorphism preserving
fibers and orientations.

We use the following notation with respect to an oriented fibered
solid torus $V(\beta/\alpha)$: M will be a suitable oriented meridian of the
solid torus $V(\beta/\alpha)$; H will denote an oriented fiber on $\partial V(\beta/\alpha)$;
Q will be an oriented single curve on $\partial V(\beta/\alpha)$ which "forms a basis"
with H, and such that $Q \cdot H = +1$ (on $\partial V(\beta/\alpha)$) and $M \sim \alpha Q + \beta H$ (on
$\partial V(\beta/\alpha)$), where we assume that α, β are relatively prime. To obtain
such a Q is an easy exercise: one starts with some Q' with
$Q' \cdot H = 1$ (on $\partial V(\beta/\alpha)$) and rectifies it by multiples of H, until the
condition $M \sim \alpha Q + \beta H$ on $\partial V(\beta/\alpha)$ is satisfied. For instance, in the
case of Fig. 1, we can start with

$$Q' \sim -2M + 3L$$
$$H \sim -5M + 7L \quad \text{on } \partial V(4/7)$$

which is equivalent to

$$M \sim 7Q' - 3H$$
$$L \sim 5Q' - 2H \quad \text{on } \partial V(4/7)$$

and take $Q \sim Q' - H = 3M - 4L$ to get $M \sim 7Q + 4H$ (Fig. 4).

An orientable 3-manifold admitting an S^1-action such that every
element different from the identity moves something (i.e. an
effective action), and such that the isotropy subgroup of a point
is never S^1, is a Seifert manifold, where the fibers are the orbits
of the action.

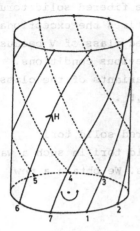

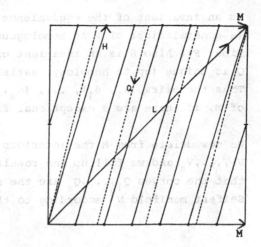

Fig. 4. The fiber H, and Q: $\beta/\alpha = 4/7$

Exercise 1. The S^1-action on

$$S^3 = \{(z_1, z_2) \in \mathbb{C}^2 : z_1\bar{z}_1 + z_2\bar{z}_2 = 1\}$$

given by

$$(t, z_1, z_2) \mapsto (e^{\alpha t i}z_1, e^{\beta t i}z_2) ,$$

where $0 < \beta < \alpha$ *and* g.c.d.$(\alpha, \beta) = 1$*, defines a Seifert structure on* S^3*. Determine the exceptional fibers (at most two) and show that any general fiber is a torus knot.*

4.2 Invariants

We now describe a set of invariants which determine the equivalence class of an oriented Seifert manifold.

Let F be an exceptional fiber of the oriented Seifert manifold M^3. Let V be a fibered solid torus which is a tubular neighbourhood of F and assume V has the orientation induced by that of M^3. By the last section, we can find oriented curves H and Q on ∂V such that H is a (general) fiber of M^3, $Q \cdot H = 1$ (on ∂V), and $M \sim \alpha Q + \beta H$, $0 < \beta < \alpha$, where M is a meridian of V suitably oriented. Clearly α

is an invariant of the equivalence class of the fibered solid torus
(a general fiber of V is homologous in V to α times the exceptional
fiber F). Also β is an invariant of the oriented class of V because
Q is unique (up to homology) satisfying the previous conditions.
Thus the pairs $(\alpha_1, \beta_1), \ldots, (\alpha_s, \beta_s)$ are invariants of the class
of M, if there are s exceptional fibers $F_1, \ldots, F_s$.

We now delete from M the interiors of the fibered solid tori
$V_1, \ldots, V_s$ and we fill up the result with s solid tori in such a way
that the curves $Q_1, \ldots, Q_s$ are the new meridians. We obtain a new
Seifert manifold N^3 according to the

*Exercise 1. Let V be a solid torus and H a simple closed curve on
∂V which is not a meridian. Then V can be fibered so that it becomes
a fibered solid torus with fiber H. The equivalence class of this
fibration is unique.*

The new Seifert manifold N^3 has no exceptional fibers, because the
meridian of the new V_i is Q_i which is homologous to $Q_i + O \cdot H_i$ (on
∂V_i). Thus N^3 is just a (locally trivial) S^1-bundle with an
orientable base F_g, or non-orientable N_k, and Euler number e. It is
clear that the character of orientability of the base, its genus and
e, are invariants of the equivalence class of M. We will see in the
next section that these invariants, together with $\alpha_1/\beta_1, \ldots, \alpha_s/\beta_s$,
determine the class of M^3 by actually constructing M, given these
data. We represent M by the notation

$$(Oog \mid -e; \ (\alpha_1, \beta_1), \ldots, (\alpha_s, \beta_s))$$

or by

$$(Onk \mid -e; \ (\alpha_1, \beta_1), \ldots, (\alpha_s, \beta_s)) \ ,$$

where the first "O" means that M is oriented; "o" or "n" means that
the base is orientable or non orientable, respectively. The reason
for the minus-sign in e will be clarified in the next section.

*Exercise 2. Check that the base of N^3 is a closed surface (obtained
by collapsing fibers).*

Exercise 3. A Seifert manifold whose fibers are orbits of some S^1-*action has orientable base.*

Hint.- If the base contains a Möbius band, the fibration over it contains a fibered Klein bottle.

4.3 Constructing the manifold from the invariants

First, we construct $(Oog|0)$. The case $g = 0$, i.e. $S^1 \times S^2$, is obtained by 0-surgery on the trivial knot (Fig. 5). We assume that S^3 is oriented, and this orientation is given by a right-handed screw in our pictures. The reader can refer to [R] for the concept and notations of Dehn-surgery.

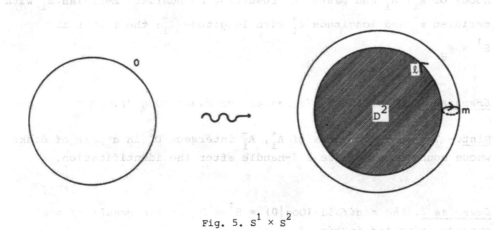

Fig. 5. $S^1 \times S^2$

Exercise 1. Check that the manifold so defined is $S^1 \times S^2$. *Draw the fibers.*

Hint.- The disk D^2 of Fig. 5, together with the one bounded by ℓ after the 0-surgery, form the S^2-factor.

The manifold $(Oog|0) = S^1 \times F_g$ can be described as follows (Fig. 6). The curves $A_1^+, A_1^-, \ldots, A_g^+, A_g^-$ are fibers of $S^1 \times S^2$. Delete neighbour-

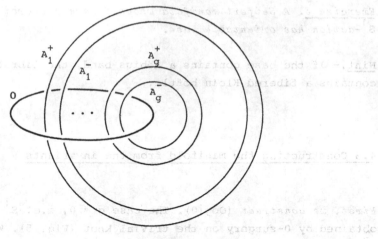

Fig. 6. $S^1 \times F_g$

hoods of A_i^+, A_i^- and paste the resulting boundaries: meridian m_i^+ with meridian m_i^- and longitude ℓ_i^+ with longitude ℓ_i^-: the result is $S^1 \times F_g$.

Exercise 2. Check that this process produces $(Oog|0)$.

Hint.- The neighbourhoods of A_i^+, A_i^- intersect D^2 in a pair of disks whose boundaries produce a 1-handle after the identification.

Exercise 3. The manifold $(Oog|0) = S^1 \times F_g$ is the result of the surgery indicated in Fig. 7.

Hint.- The identification between the boundaries of $N(A_i^+)$, $N(A_i^-)$ is interpreted as follows. Take $S^1 \times B^3$ with $S^1 \times S^2 = \partial(S^1 \times B^3)$, and paste $S^1 \times [0, 1] \times D^2$ to $S^1 \times S^2$ so that $S^1 \times \{0, 1\} \times D^2$ is identified with $N(A_i^+) \cup N(A_i^-)$. The resulting boundary is $S^1 \times F_g$. We now paste $S^1 \times [0, 1] \times D^2$ in two steps. First we paste $E_+^1 \times [0, 1] \times D^2$ (E_+^1 = northern hemisphere of S^1) and this introduces a circle endowed with zero. The second step is pasting $E_-^1 \times [0, 1] \times D^2$ (Fig. 8).

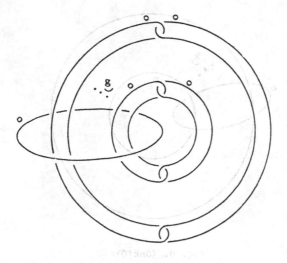

Fig. 7. A different representation of $S^1 \times F_g$. For $g = 1$ we have the Borromean rings

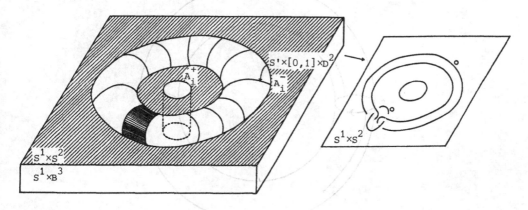

Fig. 8. Pasting $S^1 \times [0, 1] \times D^2$

We now construct $(Onk|0)$. This manifold is obtained as follows (Fig. 9). The curves $B_1, \ldots, B_k$ are fibers of $S^1 \times S^2$. Delete neighbourhoods $N(B_i)$, $i=1,\ldots,k$, and identify $\partial N(B_i)$ with itself as indicated in Fig. 10, i.e. $\partial N_i(B_i)$ can be thought of as $S^1 \times S^1$ so that $S^1 \times 0$ is a meridian of $N_i(B_i)$ and $0 \times S^1$ is a longitude of $\partial N_i(B_i)$ (null-homologous in $S^3 - B_i$) and identify (α, β) with $(a\alpha, c\beta)$, where $a: S^1 \to S^1$ is a 180° rotation and c is conjugation in $S^1 \subset \mathbb{C}$.

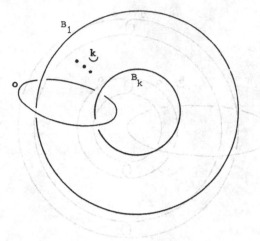

Fig. 9. (Onk|0)

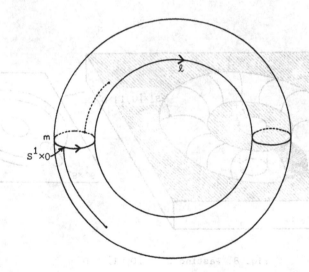

Fig. 10. Identifications in $\partial N(B_i)$

Exercise 4. *Check that this produces* (Onk|0).

<u>Hint</u>.- Compare with 1.7. The disk D^2 of Fig. 5 is k times punctured by the $N(B_i)$'s. The identifications convert it to a non-orientable surface of genus k. Thus the resulting manifold has a section.

Exercise 5. *Show that* (Onk|0) *is also given by the surgery of Fig. 11.*

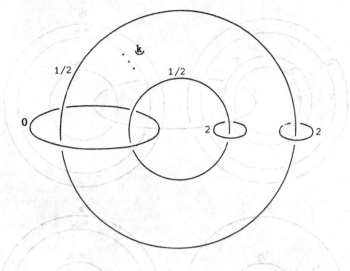

Fig. 11. (Onk|0)

Finally *we construct* (Oog|b; $(\alpha_1, \beta_1),\ldots,(\alpha_s, \beta_s)$) and
(Onk|b; $(\alpha_1, \beta_1),\ldots,(\alpha_s, \beta_s)$). Take s+1 fibers of (Oog|0) or
(Onk|0) and perform surgeries 1/b, $\alpha_1/\beta_1,\ldots,\alpha_s/\beta_s$ (Fig. 12). To
see this, note that if in the boundary of a neighbourhood N(F) of
a general fiber F, we take the basis Q, H as in 1.2, then Q·H = +1
(on ∂N(F)).

Note also that, for the definition of the Euler number in 1.2, we
took Q with the orientation induced by the punctured section; thus
Q has here the opposite orientation. That is why we take b = -e.

4.4 Change of orientation and normalization

Consider $<O_{og}^{nk}|b; (\alpha_1, \beta_1),\ldots,(\alpha_s,\beta_s)>$, with $\alpha_i \neq 0$, and where the
angle brackets mean that the conditions $0 < \beta_i < \alpha_i$ are not
necessarily satisfied. Realizing the surgeries of Fig. 12, using the
values 1/b, $\alpha_1/\beta_1,\ldots,\alpha_s/\beta_s$, it is clear that we obtain a Seifert
manifold; *normalization*, in this context, means finding the Seifert
invariants of that manifold. The following geometric observation is
enough to solve this little problem:

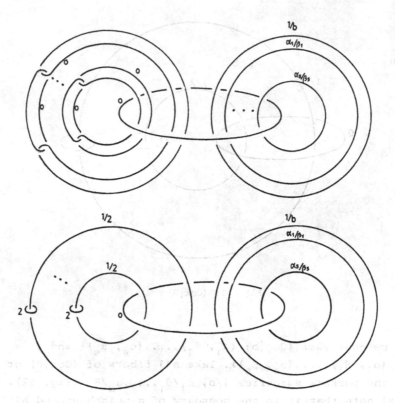

Fig. 12. Seifert manifolds $(Oog|b; (\alpha_1, \beta_1),...,(\alpha_s, \beta_s))$ and $(Onk|b; (\alpha_1, \beta_1),...,(\alpha_s, \beta_s))$

Observation.- Let us consider two parallel curves in a solid torus V, with surgery instructions (1, b) and (α, β). Then, without changing the meridian of the resulting solid torus V, we can exchange the surgery instructions for (1, b+n), (α, $\beta-\alpha n$) respectively (compare [Wa]).

To see this (Fig. 13) take the annulus A, cut along it and shift one side n complete turns, then paste back. This defines a fiber-preserving homeomorphism. Thus $m_1 + b\ell_1$ becomes $m_1 + (n + b)\ell_1$ and $\alpha m_2 + \beta\ell_2$ becomes $\alpha m_2 + (\beta - \alpha n)\ell_2$.

Example. $<On1|-2; (1, -3), (3, -1), (5,6)> = (On1|-5; (3,2), (5,1))$.

If we change the orientation of the manifold, the surgery instructions of Fig. 12 change to

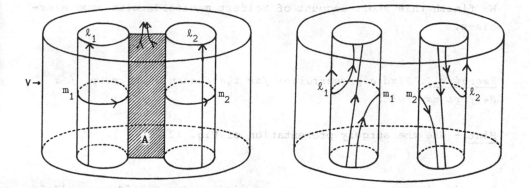

Fig. 13. Normalization (n = 2)

$$<O_{og}^{nk}|-b;\ (\alpha_1,\ -\beta_1),\ldots,(\alpha_s,\ -\beta_s)> = (O_{og}^{nk}|-s-b;\ (\alpha_1,\ \alpha_1-\beta_1),\ldots$$

$$\ldots,\ (\alpha_s,\ \alpha_s-\beta_s)) .$$

For instance

$$-(OoO|-1;\ (2,1),(3,1),(5,1)) = (OoO|-2;\ (2,1),(3,2),(5,4)),$$

thus this manifold does not have an orientation-reversing homeomorphism preserving fibers (even without preserving fibers!).

Some people (e.g. in [Th2], [Sc]) use the notation

$$(g;\ e_0;\ \beta_1/\alpha_1,\ldots,\beta_s/\alpha_s),\quad \text{where } g \in \mathbb{Z},$$

$$e_0 = -(b + \frac{\beta_1}{\alpha_1} + \ldots + \frac{\beta_s}{\alpha_s}) ,$$

and they called e_0 (rather appropriately (see Appendix A.4)) *the rational Euler number of the Seifert fibration*, where $1 < \alpha_1 \le \ldots \le \alpha_m$ and $\beta_i/\alpha_i \in (\mathbb{Q}/\mathbb{Z})^*$ (here $\mathbb{Q}$ is the field of rational numbers and $(\mathbb{Q}/\mathbb{Z})^*$ is the group $\mathbb{Q}/\mathbb{Z}$ minus zero). This is done so that e_0 does not depend on the representatives of β_i/α_i. Thus, for instance, $<On1\ -2;\ (1,\ -3),\ (3,\ -1),\ (5,6)> = (On1\ -5;\ (3,2),(5,1))$ is, in the new notation, $(-1;\ \frac{62}{15};\ -1/3,\ 6/5) = (-1;\ \frac{62}{15};\ 2/3,\ 1/5)$.

We finish this short account of Seifert manifolds with some exercises.

Exercise 1. Find a presentation for the fundamental group of a
Seifert manifold.

Hint.- Use the surgery presentation of Fig. 12.

Exercise 2. Determine necessary conditions for a Seifert manifold
to have a finite fundamental group.

Hint.- Since the base is a quotient of M, it must be S^2 or $\mathbb{R}P^2$.
After killing H and obtaining a presentation

$$|Q_1,\ldots, Q_n: Q_1\ldots Q_n = Q_1^{\alpha_1} = \ldots = Q_n^{\alpha_n} = 1|\quad,$$

interpret this group as generated by a polygon in the euclidean,
spherical or hyperbolic plane (compare 5.1). Excluding the euclidean
and hyperbolic cases (not compact), deduce that there are at most
three exceptional fibers if the base is S^2 (or just one if it is
$\mathbb{R}P^2$: take the covering $S^2 \to \mathbb{R}P^2$) and use Chapter Three.

Exercise 3. Find the Seifert fibrations with $\pi_1 = 1$ and check that
they are the fibrations for S^3 indicated in Exercise 1.1 (thus the
Poincaré conjecture is true for Seifert manifolds).

Exercise 4. Find necessary and sufficient conditions for a Seifert
manifold to be a homology sphere.

Exercise 5. Find the Seifert fibrations of the lens space $L(3, 1)$.

Exercise 6. Take the pullback of the Hopf fibration $S^3 \to S^2$, under
$z \mapsto z^m$, where $z \in \mathbb{C} + \infty \cong S^2$, and identify the fibration thus
obtained.

4.5 The manifolds of euclidean tessellations as Seifert manifolds

The manifolds of euclidean tessellations are Seifert manifolds. We
will find the invariants geometrically (i.e. from the tessellation
itself), for the manifold M(S442) of square tiles of side 1 (see
2.1). Remember that this is an oriented manifold because the charts
defined in 2.1 define local orientations on M(S442), inherited from
the usual orientation of $\mathbb{R}^2 \times \mathbb{R}$, and the changes of charts are
given by orientation preserving diffeomorphisms. Alternatively
M(S442) is the quotient of the oriented manifold $ST(E^2)$ under the
group of orientation-preserving isometries S442.

We now observe that M(S442) has an S^1-action (in fact, infinitely
many, all of them conjugated by translations): Take any point $O \in E^2$
(fixed from now on) and consider the rotation $r_t: E^2 \to E^2$ of angle t,
around O. Define the S^1-action by

$$\varphi_O: S^1 \times M(S442) \to M(S442)$$

$$(t, c) \mapsto r_t\, c$$

Exercise 1. Show that this action is differentiable and effective,
using the description of M(S442) by charts.

Exercise 2. φ_O, φ_P are conjugate, for every $P \in \mathbb{R}^2$.

Now we obtain the *exceptional fibers* for the Seifert structure given
in M(S442) by this action.

The "exceptional positions" of the tessellations with respect to O
define the points of the exceptional fibers. An "exceptional position" is one which is "not generic" in the sense that the smallest
perturbation changes the quality of the position. Thus, for
instance, the three positions of Fig. 14 are exceptional; any other
position is general ("generic"). In fact, the tessellation c, of
Fig. 14, produces an orbit which closes itself after a quarter of
a turn, and defines an exceptional orbit of *multiplicity* $\alpha = 4$. Also
c' and c" define exceptional orbits of multiplicities 4 and 2 re-

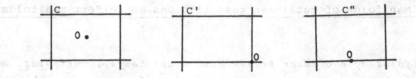

Fig. 14. These non-generic positions generate the exceptional fibers

spectively. Thus there are precisely three exceptional fibers of multiplicities α = 4, 4, 2, because for tessellations in generic positions the orbit closes after a complete turn, giving rise to general fibers.

To obtain the value of β for the exceptional fiber F we make use of the following

Rule.- (i) Find a meridian m of a tubular neighbourhood N(F) and orient it so that m followed by the orientation of an orbit S^1x, $x \in m$, gives the orientation of $\partial N(F)$ (Figs. 1 and 15).

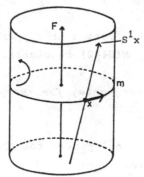

Fig. 15. Orientation of m

(ii) Number the points of $m \cap S^1x$, starting with $x = 0$ and continuing with $1, 2, \ldots, \alpha-1$ following the orientation of m (Figs. 1 and 16).

(iii) The number which follows 0 on S^1x is β^*. Then β is calculated for $\beta\beta^* \equiv 1 \mod \alpha$. The rule really means that β^*/α is the *holonomy*

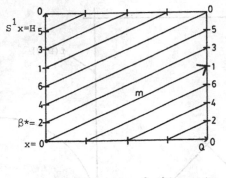

Fig. 16. Computing β (β/α = 4/7)

of the circle-foliation in the torus V(β/α). In fact, what we have
seen in the rule is that taking a general fiber $S^1 x$, close to the
exceptional fiber F, and a disk (with boundary m) transverse to F,
the first return of the general fiber, starting in D and returning
to D, is a turn of angle $2\pi\beta*/\alpha$ (compare [BS]).

Now we apply this rule to the fiber $F = S^1 c$, where c is a tessellation (parallel to the coordinate axis) having a vertex at O. Take
a neighbourhood $f(V \times A)$ of $c = f(O,O)$ in $M(S442)$, where V is a
disk in $\mathbb{R}^2$ of center O and radius $\frac{1}{4}$, A is $\left[-\frac{\pi}{4}, \frac{\pi}{4}\right] \subset \mathbb{R}^1$ and
$f: \mathbb{R}^2 \times \mathbb{R} \rightarrow M(S442)$ is the parametrization used in 2.1. We take
the set $S^1 f(V \times 0)$ as a neighbourhood N(F) of F. We can see in Fig.
17 that $f(V \times 0)$ is transverse to the fibers of the S^1-action. In
fact

$$f(V \times A) \subset S^1 f(V \times 0) \quad ,$$

and the S^1-action moves $f(V \times 0)$ turning it by the angle $t \in S^1$ and
lifting it, at the same time, to the level t. This shows that
$f(V \times 0)$ is a meridian disk of

$$S^1 f(V \times 0) = N(F) .$$

Hence, m *consists of tessellations parallel to c with vertices in*
∂V (Fig. 18). The orientation of $f(V \times A)$ is the orientation of V
followed by the orientation of A, and since the orientations of
$S^1 c$ and $f(O \times A)$ coincide (Fig. 17) we can identify the oriented m
with ∂V. We have thus accomplished part (i) of the rule.

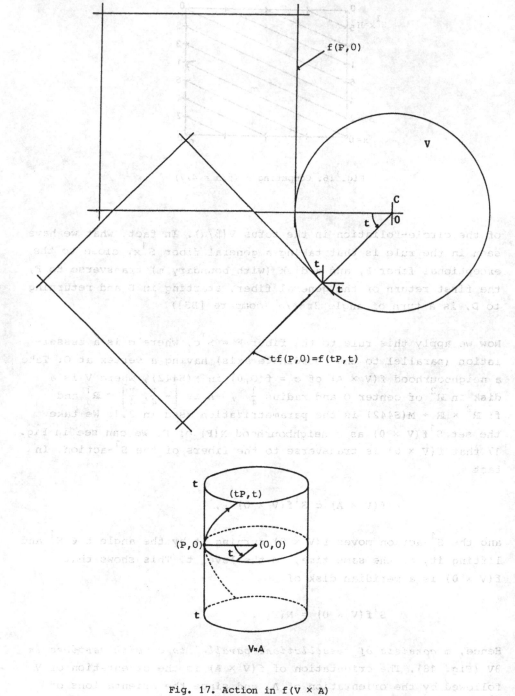

Fig. 17. Action in f(V × A)

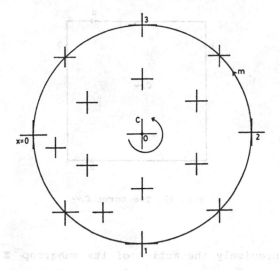

Fig. 18. Meridian disk in N(F)

To compute β^*, enumerate the tessellations of $m \cap S^1x$ (Fig. 18) following the orientation of m. The tessellation following x = 0 in S^1x, is the one with number 1, hence $\beta^* = 1$. Thus $\beta = 1$.

The same argument, applied to the other cases, implies that the exceptional fibers have the invariants (4, 1), (4, 1) and (2, 1).

We now *calculate the base* of the Seifert fibration. Let C be a square of side 1 and contained in $\mathbb{R}^2$, and let T be the subspace of M(S442) formed by tessellations parallel to the sides of C (and which necessarily have some vertex in C). The map g:C → T which sends v ∈ C to the tessellation Cv ∈ T with vertex in v, is continuous and factors through the torus C/~ of Fig. 19. Thus T is a torus. Moreover, since for every tessellation x of M(S442),

$$S^1x \cap T \neq \emptyset \quad,$$

we see that

$$S^1T = M(S442) \quad.$$

Hence the base of the fibration will be T/S^1. But for each quarter of turn, every tessellation becomes parallel to itself, hence the action

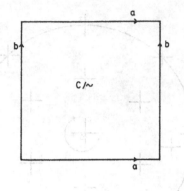

Fig. 19. The torus C/~

of S^1 on T is precisely the action of its subgroup $\mathbb{Z}_4 \subset S^1$. A fundamental domain for the action of $\mathbb{Z}_4$ on T is any of the four little squares of Fig. 20. Clearly $T/\mathbb{Z}_4 \cong S^2$.

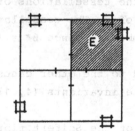

Fig. 20. Fundamental domain for the action of $\mathbb{Z}_4$ in T

Finally, *we will calculate b*. We use the fact that there is a torus $T^2 \subset M$ cutting the orbits of the S^1-action transversally. Take $\mathbb{Z}_4 \subset S^1$. Then $\mathbb{Z}_4$ acts as a group of diffeomorphisms on M(S442) and

$$M(S442) \to M(S442)/\mathbb{Z}_4$$

is a branched covering. Since $\mathbb{Z}_4$ preserves fibers of the Seifert fibration, it follows that $M(S442)/\mathbb{Z}_4$ is a Seifert manifold: its fibers are quotients under $\mathbb{Z}_4$ of fibers of M(S442). We calculate the invariants of $M(S442)/\mathbb{Z}_4$. The base is the same as the base of M(S442), i.e. S^2. But $M(S442)/\mathbb{Z}_4$ *has no exceptional fibers*. In fact, let us see how to realize the quotient under $\mathbb{Z}_4$ in a tubular neighbourhood of an exceptional fiber of M(S442), for instance of type (4, 1). As is shown in Fig. 21, the fiber downstairs has a tubular neighbourhood with meridian m' = Q' + H' where H', Q' are images of H and Q: In general (α, β) gives $(\alpha, h\beta)$, under the action of $\mathbb{Z}_h \leq S^1$, in Fig. 21. Thus

$$M(S442)/\mathbb{Z}_4 = \langle Oo0 | 4b; \ (1,1),(1,1),(1,2) \rangle = (Oo0 | 4b + 4) \ .$$

But the torus T embeds in M(S442) and is transverse to the fibers, and we have seen before that $T/\mathbb{Z}_4 \cong S^2$ embeds in $M(S442)/\mathbb{Z}_4$ and is transverse to the fibers. Hence $M(S442)/\mathbb{Z}_4$ has a section; hence 4b + 4 = 0. Thus b = -1 and

$$M(S442) = (Oo0 | -1; \ (4,1),(4,1),(2,1)) \ .$$

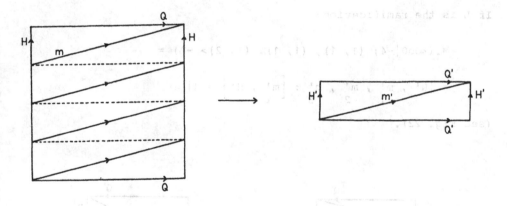

Fig. 21. Quotient in the neighbourhood of an exceptional fiber

Exercise 3. Show that the manifolds of euclidean tessellations are

        M(T)   = (Oo1|0)
        M(S2222) = (Oo0|-2; (2, 1), (2, 1), (2, 1), (2, 1))
        M(S442) = (Oo0|-1; (4, 1), (4, 1), (2, 1))
        M(S333) = (Oo0|-1; (3, 1), (3, 1), (3, 1))
        M(S632) = (Oo0|-1; (6, 1), (3, 1), (2, 1))

Check that the rational Euler class is zero.

Exercise 4. Show that $S^1 \times F_1$ *is a branched covering over the manifolds of euclidean tessellations.*

Exercise 5. We have seen that the manifolds of euclidean tessellations are branched coverings of $S^1 \times S^2$. Find the monodromies.

<u>Hint</u>.- For instance

$$\pi_1(<OoO|-4;\ (1,\ 1),\ (1,\ 1),\ (1,\ 2)>) =$$

$$= |Q'_0,\ Q'_1,\ Q'_2,\ Q'_3,\ H' : \left[Q'_i,\ H'\right] = 1,\ Q'_0 H'^{-4} = Q'_1 H' =$$

$$= Q'_2 H' = Q'_3 H'^2 = 1| = |Q'_0,\ H',\ m'_1,\ m'_2,\ m'_3 : \left[Q'_0,\ H'\right] =$$

$$= \left[m'_i, H'\right] = 1,\ Q'_0 H'^{-4} = 1,\ m'_1 = m'_2 = m'_3 = 1| \quad.$$

If L is the ramification:

$$\pi_1(<OoO|-4;\ (1,\ 1),\ (1,\ 1),\ (1,\ 2)> -L) =$$

$$= |H',\ m'_1,\ m'_2,\ m'_3 : \left[m'_i,\ H'\right] = 1|$$

(see Fig. 22).

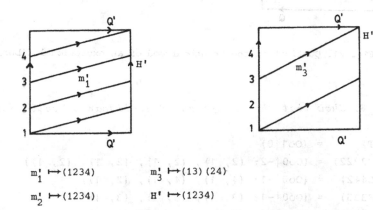

$$m'_1 \longmapsto (1234) \qquad m'_3 \longmapsto (13)(24)$$

$$m'_2 \longmapsto (1234) \qquad H' \longmapsto (1234)$$

Fig. 22. Monodromy for $(OoO|-1;\ (4,\ 1),(4,\ 1),(2,\ 1)) \xrightarrow{4:1} S^1 \times S^2$

Exercise 6. *The manifolds of euclidean tessellations are of the form*
$E(2)/\Gamma$ *which is (orientation-reversing) homeomorphic to* $\Gamma\backslash E(2)$. *Show*
that the S^1-*action on the last manifold is the action induced by*

$$(E^2 \times S^1) \times S^1 \longrightarrow E^2 \times S^1$$

$$(v, \alpha) \longmapsto (v, \alpha - t)$$

under the identification $E(2) \cong \mathbb{R}^2 \times S^1$ *(Fig. 23).*

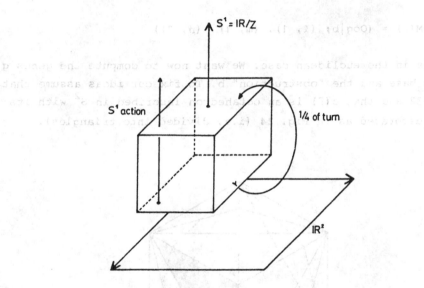

Fig. 23. S^1-action on $(OoO|-1;\ (4,\ 1),\ (4,\ 1),(2,\ 1))$

<u>Hint.</u>-

$$(v,\ \alpha)\,\Gamma \longleftrightarrow \Gamma(-ve^{-i\alpha},\ -\alpha)$$

$$\downarrow t$$

$$(ve^{it},\ \alpha + t)\,\Gamma \longleftrightarrow \Gamma(-ve^{it}e^{-i(\alpha + t)},\ -\alpha-t).$$

4.6 The manifolds of spherical tessellations as Seifert manifolds

Let Γ be a finite subgroup of SO(3) and consider the manifold $M(\Gamma)$ of the canonical tessellation $c(\Gamma)$. As in the euclidean case, the rotation around a diameter A of S^2 defines an S^1-action on $M(\Gamma)$. Thus $M(\Gamma)$ is a Seifert manifold and, in fact, in as many ways as possible selections of the diameter A. All of these Seifert structures of $M(\Gamma)$ are conjugate by the rotations which change the axis.

If Γ is the group Sℓmn then

$$M(\Gamma) = (Oog|b; (\ell, 1), (m, 1), (n, 1))$$

just as in the euclidean case. We want now to compute the genus g of the base and the "obstruction" b. To fix our ideas assume that Γ = S432 and that $c(\Gamma)$ is an octahedron inscribed in S^2 with its faces decorated as in Fig. 24 (i.e. divided into triangles).

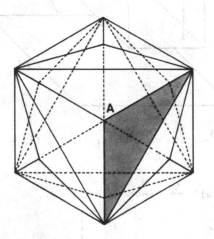

Fig. 24. A decorated octahedron

Exercise 1. Show that the base of the Seifert structure of $M(\Gamma)$ is the orbifold which is the quotient of S^2 under the group Γ.

Hint.- The shaded triangle of Fig. 24 defines a subset of $M(\Gamma)$, namely, the set of octahedra having a vertex in that triangle. The orbits of the elements of this set under the S^1-action cover $M(\Gamma)$.

The last exercise, which should be compared to an analogous calculation in the euclidean case, shows that $g = 0$.

Now we "asymmetrize" the octahedron of Fig. 24. This can be done, for instance, by colouring the triangles of its boundary with different colours. The group of symmetry Γ' of that asymetric octahedron is the identity, and the manifold $M(\Gamma')$ of the tessellation of Γ' is SO(3). Thus we have our familiar covering

$$f: M(\Gamma') \to M(\Gamma)$$

with 24 sheets (recall the table in 3.8). The manifold

$$M(\Gamma') = SO(3)$$

has the S^1-action induced by rotation around A, and f preserves fibers. Since the Seifert fibration induced by rotation around A in SO(3) does not have exceptional fibers, it defines an S^1-bundle structure on SO(3) and the Euler number must be 2 (see 1.3). Thus the Seifert structure induced in SO(3) by rotation around A is $(Oo0|2)$.

Exercise 2. *Deduce that the base of the Seifert structure of* $M(\Gamma)$ *has genus* $g = 0$ *by using the fact that* f *is fiber-preserving.*

Exercise 3. SO(3) *has the* S^1-*bundle structure coming from its identification with* $ST(S^2)$, *and it also has infinitely many* S^1-*bundle structures coming from rotations around axes like* A. *Construct the equivalences among these structures.*

In the covering

$$f:M(\Gamma') = (Oo0|2) \to M(\Gamma) = (Oo0|b;\ (4,1),\ (3,1),\ (2,1))$$

the preimage of a general fiber consists of 24 fibers; the preimage of an exceptional fiber of order 4, F_4, consists of 6 fibers of $M(\Gamma')$, each one covering F_4 with degree 4: this is because, if $c(\Gamma)$ has a vertex in common with A, $c(\Gamma)$ comes back to itself after $\frac{2\pi}{4}$ of turn, but $c(\Gamma')$ comes back after 2π. Similarly, the preimages of F_3 and F_2 consist of 8 and 12 fibers respectively.

Using Exercise 1 of 4.4 we can obtain $\pi_1(M(S432))$ from the surgery presentation of Fig. 25 (compare Fig. 12). The presentation obtained is

$$|Q_0, Q_1, Q_2, Q_3, H : \left[Q_i, H\right] = 1, \ i=0, \ 1, \ 2, \ 3; \ Q_0 \ H^b = Q_1^4 \ H =$$

$$= Q_2^3 \ H = Q_3^2 \ H = Q_0 \ Q_1 \ Q_2 \ Q_3 = 1|.$$

Killing the center (generated by H) we obtain a presentation for the group Γ = S432, and the composition

$$w: \ \pi_1(M(S432)) \ \rightarrow \ \Gamma = S432 \ \hookrightarrow \ S_{24}$$

where the last map is the inclusion of S_4 in S_{24}, defines the monodromy of the covering f (the kernel of w is generated by H and is $\mathbb{Z}_2$; compare 3.8). Now the covering f is very easy to construct. In Fig. 26 we see how the boundary of a tubular neighbourhood of the fiber F_4 lifts. The covering has the surgery presentation shown in Fig. 27, i.e. it is the Seifert manifold

$$<Oo0|24b; \ (1,1),..\overset{6}{..},(1,1),(1,1),..\overset{8}{..},(1,1),(1,1),..\overset{12}{..},(1,1)> =$$

$$= (Oo0|24b + 26) \ .$$

Since this manifold is $(Oo0|2)$, we must have b = -1.

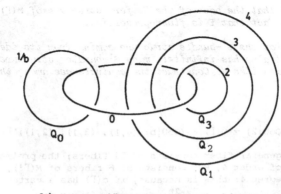

Qi's are meridians

Fig. 25. The manifold M(S432)

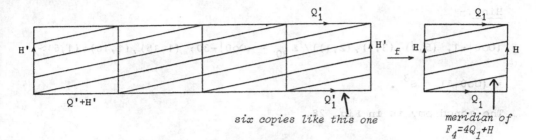

six copies like this one

meridian of $F_4 = 4Q_1 + H$

Fig. 26. Preimage of tube around F_4

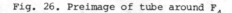

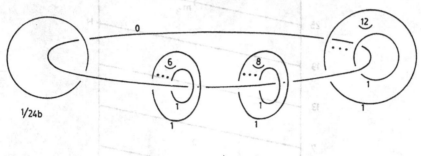

Fig. 27. $(O o 0 | 2)$

In general, if θ is the order of Γ we must have

$$2 = \theta(b + \frac{1}{\ell} + \frac{1}{m} + \frac{1}{n}) \ .$$

For

$$(\ell, \ m, \ n) \in \{(m, \ 2, \ 2), \ (3, \ 3, \ 2), \ (4, \ 3, \ 2), \ (5, \ 3, \ 2)\}$$

we have $b = -1$. For $(m, m, 1)$, $b = 0$. Thus we finally have

$$
\begin{aligned}
M(Smm) &= (O o 0 | 0; \ (m, \ 1), \ (m, \ 1)) \\
M(Sm22) &= (O o 0 | -1; \ (m, \ 1), \ (2, \ 1), \ (2, \ 1)) \\
M(S332) &= (O o 0 | -1; \ (3, \ 1), \ (3, \ 1), \ (2, \ 1)) \\
M(S432) &= (O o 0 | -1; \ (4, \ 1), \ (3, \ 1), \ (2, \ 1)) \\
M(S532) &= (O o 0 | -1; \ (5, \ 1), \ (3, \ 1), \ (2, \ 1)).
\end{aligned}
$$

Exercise 4. Represent the last manifolds as branched coverings of S^3 with m, 2m, 6, 12 and 30 sheets respectively, branched over 2 or 3 curves L of the Hopf fibration. Find the monodromy.

162

Hint.-

$(OoO|-1, (5,1),(3,1),(2,1))/\mathbf{Z}_{30} = <OoO|-30; (1,15),(1,10),(1,6)> =$

$= (OoO|1) = S^3$.

The monodromy is in Fig. 28.

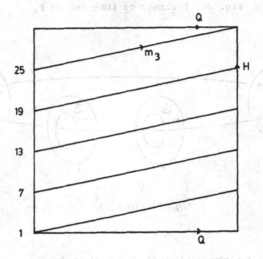

$m_3 = (1,7,13,19,25)(2,8,...) \ldots$
$m_2 = (1,11,21)(2,12,24) \ldots$
$m_1 = (1,16)(2,17) \ldots$

Fig. 28. Monodromy of $(OoO|-1; (5,1),(3,1),(2,1)) \xrightarrow{30:1} S^3$; m_i meridians of L

Exercise 5. S^1 *acts on the manifolds of euclidean and spherical tes-sellations (rotation around* $O \in E^2$ *or* S^2*). Thus* $\mathbf{Z}_m \leq S^1$ *acts also. Find the actions that are free and the corresponding quotient mani-folds. Are there new manifolds? (compare* [TS]*, page 576, and* [RV]*, page 165).*

4.7 Involutions on Seifert manifolds

We are interested in orientation-preserving involutions of Seifert manifolds. They will be of some use in 4.8. For more details see [Mo] and [BS].

But first take the fibered torus T^2 of Fig. 29. The fibers are curves
homologous to $\alpha Q + \beta H$, with $g \cdot c \cdot d(\alpha, \beta) = 1$. The quotient space of
T^2 under the involution which is rotation around A is the "fibered"
sphere S^2 (in the language of orbifolds, it is an S^1-bundle over the
1-orbifold $D^1\bar{1}\bar{1}$, i.e. an interval with silvered boundary). We have
a 2-fold covering $p: T^2 \to S^2$ branched over 4 points of S^2.

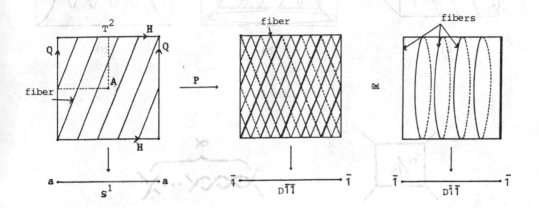

Fig. 29. $\alpha Q + \beta H = 5Q + 2H$

Consider now the *solid torus*

$$V^3 = (T^2 \times [0, 1])/\sim \quad,$$

where $\sim$ means that each fiber $\alpha Q + \beta H$ in $T^2 \times 1$ is collapsed to a
point. We can also take the 3-ball

$$B^3 = (S^2 \times [0, 1])/\sim \quad,$$

where $\sim$ means that each fiber $p(\alpha Q + \beta H)$ is collapsed to a point.
Then $p \times id$ defines the 2-fold covering

$$p': V^3 \to B^3$$

depicted in Fig. 30. The branching set is the tangle of Fig. 30c.
This tangle will be denoted by the diagram of Fig. 30d. The tangle
$\alpha/1$ is shown in Fig. 30e.

164

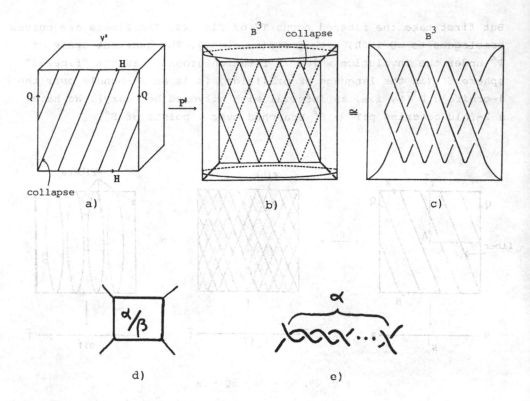

Fig. 30. p': $V^3 \to B^3$, and rational tangle α/β

In the language of orbifolds, the torus V^3 is an S^1-bundle of the
2-orbifold with boundary shown in Fig. 31a. Also B^3 is an S^1-bundle
of the 2-orbifold shown in Fig. 31b (see [BS] and A.4). In the last
case there are two types of fibers: S^1 and $D^1\overline{\overline{11}}$.

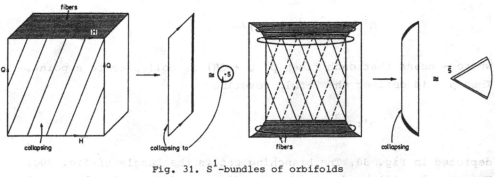

Fig. 31. S^1-bundles of orbifolds

Now we are ready to describe our involutions on the Seifert manifolds of Fig. 12. Notice that the links of Fig. 12 are symmetric with
respect to the axis A depicted in Fig. 32. Then the rotation of
180° around A defines a fiber-preserving 2-fold covering over the
manifolds shown in Fig. 33 with branching sets depicted in the same
Figure. To see this remember the definition of surgery given in 1.2
and the above considerations.

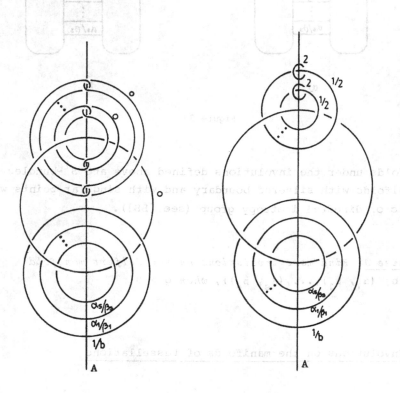

Figure 32

The manifold of Fig. 33a is

$$S^1 \times S^2 \#\ \underset{g}{\ldots..}\ \# S^1 \times S^2$$

(for g = 0 it is just S^3); the manifold of Fig. 33b is S^3.

In the language of orbifolds, we see that the Seifert manifolds are
just S^1-bundles of 2-orbifolds without boundary and with isolated
singular points with cyclic isotropy group. The quotients of Seifert

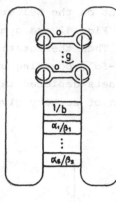

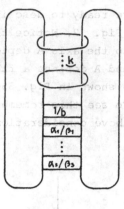

a)

b)

Figure 33

manifolds under the involutions defined above are S^1-bundles of
2-orbifolds with silvered boundary and with singular points with
cyclic or dihedral isotropy group (see [BS]).

Exercise 1. *Find more involutions on the Seifert manifold*
$(Oog|b; (\alpha_1, \beta_1),\ldots,(\alpha_s, \beta_s))$, *when* $g > 0$.

4.8 Involutions on the manifolds of tessellations

We can use our knowledge of Seifert structures on the manifolds
of euclidean or spherical tessellations to understand the involutions
defined in 2.8 and 3.15. We will take the axis E of reflection co-
incident with the center O of rotations which defines the Seifert
structure of $M(\Gamma)$.

Thus the involutions $u(D\overline{2}\overline{2}\overline{2}\overline{2})$, $u(D2\overline{2}\overline{2})$, $u(D22)$ and $u(P22)$ of the
manifold $M(S2222)$ (see Exercise 2.8.3) preserve fibers of the Seifert
structure defined by rotation around O, and reverse their orienta-
tion. Moreover, the involution $u(D\overline{2}\overline{2}\overline{2}\overline{2})$ leaves the four exceptional
fibers fixed; the involution $u(D2\overline{2}\overline{2})$ leaves only two exceptional
fibers fixed and interchanges the other two (Fig. 34); the involu-

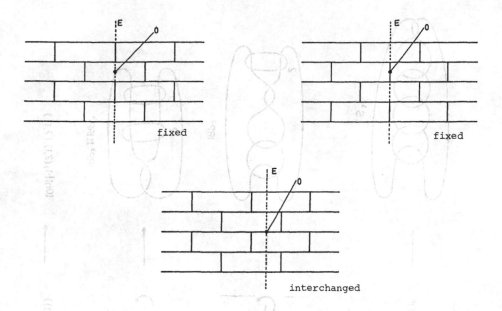

Fig. 34. Action of u(D2$\overline{2}\overline{2}$) on exceptional fibers

tions u(D22) and u(P22) interchange them in pairs; and, in case
u(P22), there is no tessellation fixed. It is then easy to recognize
these involutions: in the base of the Seifert fibration of M(S2222)
they induce the involutions giving the corresponding quotients
D$\overline{2}\overline{2}\overline{2}\overline{2}$, D2$\overline{2}\overline{2}$, D22 and P22 (Fig. 35). In the Seifert fibration itself,
the involutions, their quotients, and the branching sets are shown
in Fig. 36.

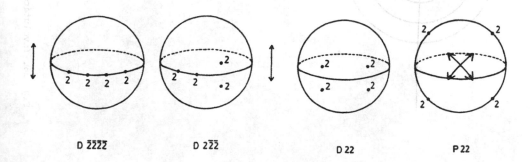

Fig. 35. The involutions on the base of the Seifert fibration

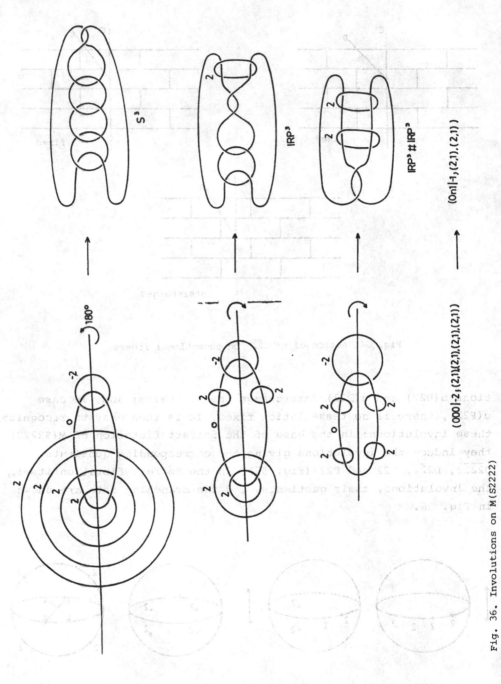

Fig. 36. Involutions on M(S2222)

Table I. Euclidean case

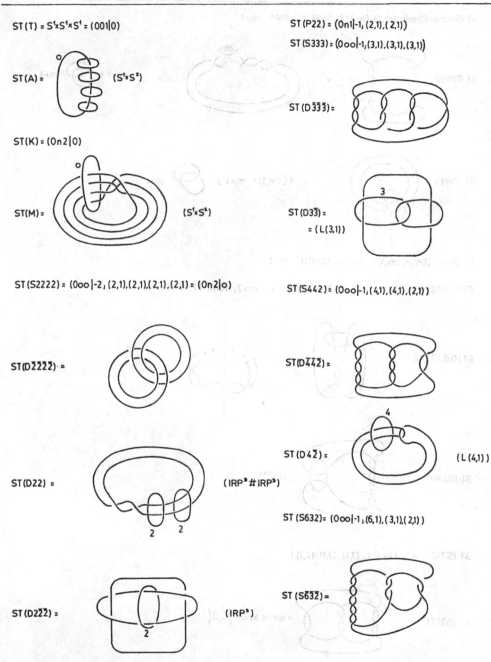

$ST(T) = S^1 \times S^1 \times S^1 = (001|0)$

$ST(A) = \qquad\qquad (S^1 \times S^2)$

$ST(K) = (0n2|0)$

$ST(M) = \qquad\qquad (S^1 \times S^2)$

$ST(S2222) = (0oo|-2; (2,1),(2,1),(2,1),(2,1) = (0n2|o)$

$ST(D\overline{2222}) =$

$ST(D22) = \qquad\qquad (IRP^3 \# IRP^3)$

$ST(D2\overline{2}) = \qquad\qquad (IRP^3)$

$ST(P22) = (0n1|-1, (2,1),(2,1))$

$ST(S333) = (0oo|-1,(3,1),(3,1),(3,1))$

$ST(D\overline{333}) =$

$ST(D3\overline{3}) =$
$= (L(3,1))$

$ST(S442) = (0oo|-1, (4,1),(4,1),(2,1))$

$ST(D\overline{442}) =$

$ST(D4\overline{2}) = \qquad\qquad (L(4,1))$

$ST(S632) = (0oo|-1; (6,1),(3,1),(2,1))$

$ST(S\overline{632}) =$

Table II. Spherical case

ST (Smm) = (0 ∞|o; (m,1), (m,1))   m⩾1 ;   IRP³   m=1

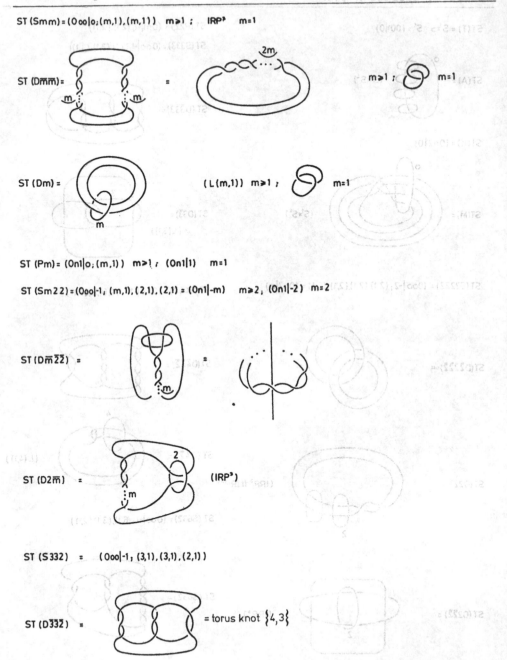

ST (Dm̄m̄) =   =   m⩾1 ;   m=1

ST (Dm) =   (L(m,1))   m⩾1 ;   m=1

ST (Pm) = (0n1|o; (m,1))   m⩾1 ;   (0n1|1)   m=1

ST (Sm22) = (0oo|-1; (m,1),(2,1),(2,1) = (0n1|-m)   m⩾2 ; (0n1|-2)   m=2

ST (Dm̄2̄2) =   =

ST (D2m̄)   =   (IRP³)

ST (S332)   =   (0oo|-1, (3,1),(3,1),(2,1))

ST (D3̄3̄2̄)   =   = torus knot {4,3}

Table II (continued)

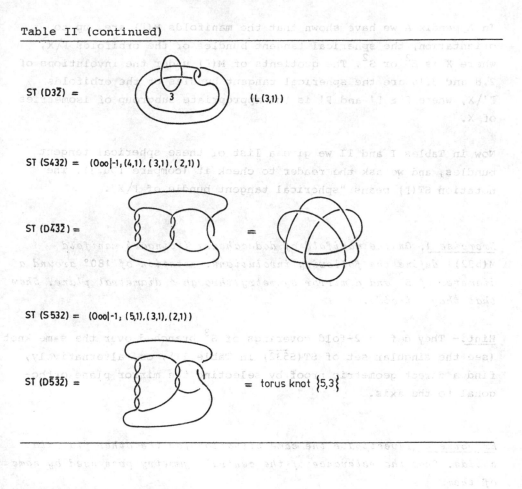

ST (D3$\bar{2}$) =   (L (3,1))

ST (S432) =   (O∞o|-1,(4,1),(3,1),(2,1))

ST (D$\bar{4}$3$\bar{2}$) =   =

ST (S 532) =   (O∞o|-1,(5,1),(3,1),(2,1))

ST (D$\bar{5}$3$\bar{2}$) =   = torus knot {5,3}

In Appendix A we have shown that the manifolds $M(\Gamma)$ are, up to orientation, the spherical tangent bundles of the orbifolds $\Gamma\backslash X$, where X is E^2 or S^2. The quotients of $M(\Gamma)$ under the involutions of 2.8 and 3.15 are the spherical tangent bundles of the orbifolds $\Gamma'\backslash X$, where $\Gamma \leq \Gamma'$ and Γ' is the appropriate subgroup of isometries of X.

Now in Tables I and II we give a list of these spherical tangent bundles, and we ask the reader to check it (compare [Du2]). The notation $ST(\Gamma)$ means "spherical tangent bundle of $\Gamma\backslash X$".

Exercise 1. On the manifold of dodecahedra (Poincaré manifold M(532)) define the following involutions: rotation of 180° around a diameter of S^2 and a mirror symmetry through a diametral plane. Show that they coincide.

Hint.- They define 2-fold coverings of S^3 branched over the same knot (see the singular set of $ST(S\overline{5}\overline{3}\overline{2})$ in Table II); or, alternatively, find a direct geometric proof by selecting the mirror plane orthogonal to the axis.

Exercise 2. Investigate the same situation for the other platonic solids. Show the relevance of the central symmetry possessed by some of them.

Hint.- See the duplicated singular sets in Table II.

Chapter Five: Manifolds of Hyperbolic Tessellations

"Cáp. CXXIII, Que sigue al ciento veinte y dos,
y trata de cosas no excusadas para claridad de esta
Historia."

*"Chapter CXXIII, Which follows the one hundred
twenty-second and deals with matters indispensable
for the clear understanding of this history."
Cervantes, Don Quixote*

In this short chapter we describe the hyperbolic tessellations and
the 3-manifolds that they define. In contrast to the euclidean and
spherical cases we will only give the Seifert invariants of the
manifolds of hyperbolic tessellations, and we will not describe them
as polyhedra with identified faces.

We have added an Appendix (Appendix B) devoted to the hyperbolic
plane, which is less familiar than the euclidean or spherical planes.

5.1 The hyperbolic tessellations

We work here with the Poincaré model of the hyperbolic plane H^2 (see
Appendix B). Let $\text{Iso}^+(H^2)$ be the group of orientation-preserving
isometries of H^2. Given two pointers (i.e. elements of $ST(H^2)$) in
H^2 there is exactly one element of $\text{Iso}^+(H^2)$ sending one to the other
(see Appendix B). Thus we can identify $\text{Iso}^+(H^2)$ with $ST(H^2)$, once a
base pointer b is fixed, by the map

$$\delta: \text{Iso}^+(H^2) \to ST(H^2)$$

$$h \mapsto dh(b)$$

This map is a diffeomorphism between the Lie group $\text{Iso}^+(H^2)$ and the differentiable manifold $\text{ST}(H^2)$. Thus δ allows us to represent isometries by pointers.

Exercise 1. *Show that* $\text{Iso}^+(H^2)$ *is diffeomorphic to the interior of a solid torus.*

Hint.- $\text{ST}(H^2)$ is a trivial bundle. Alternatively, $\text{Iso}^+(H^2)$ is $\text{PSL}(2, \mathbb{R})$ or $\text{SU}(1, 1)/(+1, -1)$ (see Appendix B).

Exercise 2. *Using* $\text{Iso}^+(H^2) \cong \text{PSL}(2, \mathbb{R})$ *show that the canonical left action of* $\text{Iso}^+(H^2)$ *on itself can be represented as a subgroup of the Lie group of projectivities of* $\mathbb{R}P^3$ *acting on the interior of a ruled hyperboloid. Describe the different types of projectivities corresponding to the three types of elements of* $\text{Iso}^+(H^2)$.

Given a subset of H^2, the subgroup Γ of $\text{Iso}^+(H^2)$ which leaves that subset invariant (as a set) is called the symmetric group of the subset. The *hyperbolic tessellations* of H^2 are the subsets of H^2 whose symmetric groups are the subgroups Γ of $\text{Iso}^+(H^2)$ acting properly-discontinuously in H^2 and having compact quotient. Thus Γ is the group of orientation-preserving isometries of the *canonical tessellation* $c(\Gamma)$, i.e. the set of pointers of H^2 which correspond to the elements of Γ. The quotient space is an orbifold $\Gamma\backslash H^2$ whose underlying metric space is a closed, orientable surface F_g. The projection $H^2 \to F_g$ is a branched covering. The branching set consists of n points of F_g whose preimages have branch-indices $\alpha_1, \ldots, \alpha_n$, respectively. The n points are the singular points of the orbifold and their isotropy groups are cyclic groups of orders $\alpha_1, \ldots, \alpha_n$. Following [BS] we will denote both Γ and the orbifold $\Gamma\backslash H^2$ by the notation $(F_g; \alpha_1, \ldots, \alpha_n)$, where $\alpha_1 \geq \alpha_2 \geq \ldots \geq \alpha_n$. In fact, these invariants characterize Γ algebraically, since it has the presentation

$$\left| A_1, B_1, \ldots, A_g, B_g, Q_1, \ldots, Q_n : [A_1, B_1] \ldots [A_g, B_g] Q_1 \ldots Q_n = Q_1^{\alpha_1} = \ldots = Q_n^{\alpha_n} = 1 \right|$$

Moreover $(F_g; \alpha_1, \ldots, \alpha_n)$ has a fundamental domain D in H^2 consisting

of a polygon whose boundary is formed of oriented geodesic arcs,
which correspond in pairs of the same length, as in Fig. 1 (a
general reference for the results of this section is [Th2]).

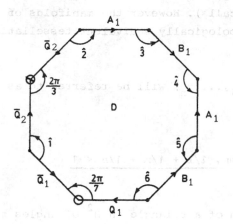

Fig. 1. Fundamental domain for $(F_1; 7,3)$; $\hat{1} + \hat{2} + \hat{3} + \hat{4} + \hat{5} + \hat{6} = 2\pi$

Each corresponding pair of arcs in ∂D defines precisely one element
of $\mathrm{Iso}^+(H^2)$ sending one arc to the other. These elements are the
generators of the above presentation of $(F_g; \alpha_1,\ldots,\alpha_n)$ where the
A_i, B_i's are hyperbolic and the Q_i's are elliptic.

Dividing D in triangles, we can compute the area of D as the sum of
the defects. Coning ∂D from an interior point we obtain $2(2g + n)$
triangles and the area of D is

$$2(2g + n)\pi - 4\pi - 2\pi(\frac{1}{\alpha_1} + \ldots + \frac{1}{\alpha_n}) =$$

$$= 2\pi(2g-2 + n -(\frac{1}{\alpha_1} + \ldots + \frac{1}{\alpha_n})) = -2\pi \chi(F_g; \alpha_1,\ldots,\alpha_n) \quad .$$

Since the area must be positive we deduce $\chi(F_g; \alpha_1,\ldots,\alpha_n) < 0$
(compare Appendix A).

Conversely (see Appendix A), given $(F_g; \alpha_1,\ldots,\alpha_n)$ with
$\chi(F_g; \alpha_1,\ldots,\alpha_n) < 0$ there exist D in H^2 and a group which realizes
the signature $(F_g; \alpha_1,\ldots,\alpha_n)$. In the construction of the domain D
there is freedom in selecting the angles in the class of vertices

adding up to 2π (Fig. 1). Thus, there are in principle different
ways of defining D, all defining the same abstract subgroup of
$Iso^{+}(H^{2})$. All of these subgroups are clearly conjugated by a homeomorphism of H^{2}. This means that the topological orbifold
$(F_{g}; \alpha_{1}, \ldots, \alpha_{n})$ might admit different hyperbolic structures (see
[Th2] for more details). However the manifolds of tessellations
defined by two topologically equivalent tessellations will be homeomorphic.

The groups $(F_{g}; \alpha_{1}, \ldots, \alpha_{n})$ will be referred to as *Fuchsian groups*.

5.2 The groups $S\ell mn$, $1/\ell + 1/m + 1/n < 1$

Let D be the union of a triangle in H^{2} of angles π/ℓ, π/m, π/n,
together with its reflection in an edge (see Fig. 2). Such a triangle
can be constructed by ruler and compass as shown in Fig. 3 (see
[Jo]). First draw ABC with angles π/m, $\pi/2 + \pi/2\ell - \pi/2n - \pi/2m$,
$\pi/2 + \pi/2n - \pi/2\ell - \pi/2m$. Then draw the isosceles triangle DBC with
equal angles $\pi/2n + \pi/2m + \pi/2\ell$. Finally draw the circles 1, 2, 3.
The interior of the circle 3 defines H^{2}. The B-angle of the *hyperbolic triangle* ABC is

$$(\pi/2 + \pi/2\ell - \pi/2n - \pi/2m) + (\pi/2n + \pi/2m + \pi/2\ell) - \pi/2 = \pi/\ell$$

and the C-angle is π/n (BC in the hyperbolic triangle is part of
circle 1).

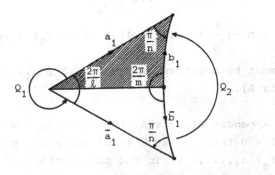

Fig. 2. Fundamental domain for $S\ell mn$

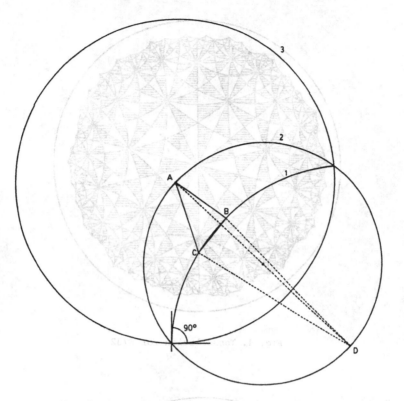

Fig. 3. Construction of a hyperbolic triangle

The isometries Q_1, Q_2 of Fig. 2 generate a group which acts properly-discontinuously on H^2, with fundamental domain D. The proof of this involves checking that the images of D under the action of the elements Q_1, Q_2 cover H^2 without overlapping. Once this is shown, a presentation for the group Sℓmn is obtained using the method of 2.10. Fig. 4 and 5 depict two tessellations of this type.

Exercise 1. Draw tessellations for F_g, $g \geq 2$

5.3 The manifolds of hyperbolic tessellations

The manifolds of tessellations in H^2 are therefore the homogeneous manifolds $\mathrm{Iso}^+(H^2)/(F_g; \alpha_1, \ldots, \alpha_n)$ where $(F_g; \alpha_1, \ldots, \alpha_n)$ is a fuchsian

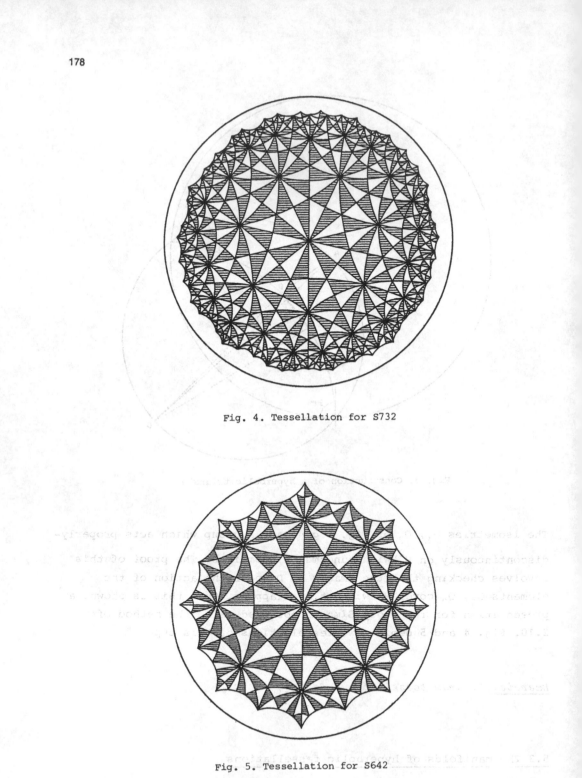

Fig. 4. Tessellation for S732

Fig. 5. Tessellation for S642

group acting canonically on the right on $\text{Iso}^+(H^2)$. It is convenient
to think also of that manifold as $M(F_g; \alpha_1,\dots,\alpha_n)$, i.e. the set of
tessellations of type $c := c(F_g; \alpha_1,\dots,\alpha_n)$ in H^2. Let

$$\eta: \text{Iso}^+(H^2)/(F_g; \alpha_1,\dots,\alpha_n) \to M(F_g; \alpha_1,\dots,\alpha_n)$$

$$\langle h \rangle \mapsto \hbar c$$

be the natural diffeomorphism.

The group $(F_g; \alpha_1,\dots,\alpha_n)$ lifts to a subgroup $(\tilde{F}_g; \alpha_1,\dots,\alpha_n)$ of the
universal covering $\widetilde{ST}(H^2)$. Since

$$\pi_1(ST(H^2)) \cong \mathbb{Z}$$

we have the exact sequence

$$1 \to \mathbb{Z} \to \pi_1(M(F_g; \alpha_1,\dots,\alpha_n)) \to (F_g; \alpha_1,\dots,\alpha_n) \to 1 \ .$$

A generator z of $\pi_1(ST(H^2))$ is a path of pointers which gives a
complete turn. Arguing as in 2.10, to find a presentation for
$\pi_1(M(F_g; \alpha_1,\dots,\alpha_n))$ it is sufficient to find the action of the relations of $(F_g; \alpha_1,\dots,\alpha_n)$ on H^2. It is clear that $Q_i^{\alpha_i}$ and
$Q_1 \dots Q_n [A_1,B_1] \dots [A_g,B_g]$ act on H^2 producing a complete turn. Hence

$$\pi_1(M(F_g; \alpha_1,\dots,\alpha_n)) = \big| z, A_1,\dots,A_g,B_1,\dots,B_g,Q_1,\dots,Q_n :$$

$$[A_1,B_1]\dots[A_g,B_g] \, Q_1\dots Q_n = Q_1^{\alpha_1} = \dots = Q_n^{\alpha_n} = z \big|$$

where z generates the center of the group and has infinite order.

5.4 The S^1-action

The diffeomorphism

$$\text{Iso}^+(H^2) \to ST(H^2)$$

$$h \mapsto dh^{-1}(b),$$

where b is some fixed pointer, induces an orientation-reversing
diffeomorphism

$$\mu: \mathrm{Iso}^+(H^2)/(F_g; \alpha_1,\ldots,\alpha_n) \to (F_g; \alpha_1,\ldots,\alpha_n)\backslash \mathrm{ST}(H^2) \quad .$$

The manifold $\mathrm{Iso}^+(H^2)/(F_g; \alpha_1,\ldots,\alpha_n)$ has been identified with
$M(F_g; \alpha_1,\ldots,\alpha_n)$ via η.

We can define an S^1-action ψ on $M(F_g; \alpha_1,\ldots,\alpha_n)$ as follows:

$$\psi: S^1 \times M(F_g; \alpha_1,\ldots,\alpha_n) \to M(F_g; \alpha_1,\ldots,\alpha_n)$$

$$(t, d) \mapsto r_t\, d \ ,$$

where r_t is a hyperbolic rotation of angle t around a fixed point
$0 \in H^2$. Via η this action corresponds to

$$\psi: S^1 \times \mathrm{Iso}^+(H^2)/(F_g; \alpha_1,\ldots,\alpha_n) \to \mathrm{Iso}^+(H^2)/(F_g; \alpha_1,\ldots,\alpha_n)$$

$$(t, <h>) \mapsto <r_t h> \ .$$

We can also define an S^1-action φ:

$$\varphi: S^1 \times (F_g; \alpha_1,\ldots,\alpha_n)\backslash \mathrm{ST}(H^2) \to (F_g; \alpha_1,\ldots,\alpha_n)\backslash \mathrm{ST}(H^2)$$

$$(t, \{v\}) \mapsto \{c_t^{-1}\, v\} \ ,$$

where $c_t: \mathrm{ST}(H^2) \to \mathrm{ST}(H^2)$ induces the identity in H^2 and rotates
each pointer of $\mathrm{ST}(H^2)$ in the angle t (Fig. 6). Notice that φ does
not preserve the orientation of the fibers of the bundle

$$(F_g; \alpha_1,\ldots,\alpha_n)\backslash \mathrm{ST}(H^2) \to (F_g; \alpha_1,\ldots,\alpha_n)\backslash H^2 \quad .$$

Now we show that *these two S^1-actions are equivalent via* μ: i.e.
the following diagram is commutative:

$$
\begin{array}{ccc}
S^1 \times \mathrm{Iso}^+(H^2)/(F_g; \alpha_1,\ldots,\alpha_n) & \xrightarrow{\;\psi\;} & \mathrm{Iso}^+(H^2)/(F_g; \alpha_1,\ldots,\alpha_n) \\
\downarrow (\mathrm{id} \times \mu) & & \downarrow \mu \\
S^1 \times (F_g; \alpha_1,\ldots,\alpha_n)\backslash \mathrm{ST}(H^2) & \xrightarrow{\;\varphi\;} & (F_g; \alpha_1,\ldots,\alpha_n)\backslash \mathrm{ST}(H^2) \quad .
\end{array}
$$

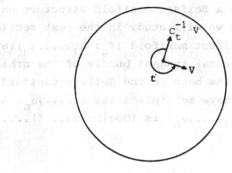

Fig. 6. S^1-action

In fact,

$$\varphi(\text{id} \times \mu)(t, \langle h \rangle) = \varphi(t, \{dh^{-1}(b)\}) = \{C_t^{-1} \, dh^{-1}(b)\} \ .$$

On the other hand

$$\mu\psi(t, \langle h \rangle) = \mu\langle r_t h \rangle = \{(dh^{-1})(dr_t^{-1})b\} \ .$$

But (see Fig. 7b) the pair $(b, dr_t^{-1}(b))$ is transformed by dh^{-1} in the pair $(dh^{-1}(b), C_t^{-1} dh^{-1}(b))$, because they span the same angle t, and h is conformal.

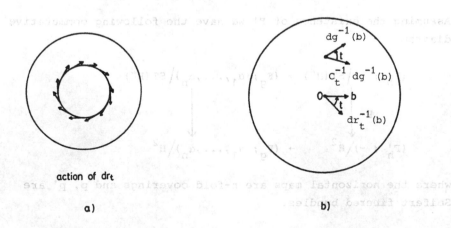

action of dr_t

a)

b)

Fig. 7. μ is equivariant

The S^1-action φ defines a Seifert manifold structure on $M(F_g; \alpha_1,\ldots,\alpha_n)$, which we will study in the next section using the orientation-reversed Seifert manifold $(F_g; \alpha_1,\ldots,\alpha_n)\backslash ST(H^2)$, which coincides with the spherical tangent bundle of the orbifold $(F_g; \alpha_1,\ldots,\alpha_n)$. Thus, the base of the Seifert manifold is F_g and the exceptional fibers have multiplicities $\alpha_1,\ldots,\alpha_n$. Arguing as in 4.5 we see that $M(F_g; \alpha_1,\ldots,\alpha_n)$ is $(Oog|b; (\alpha_1, 1),\ldots,(\alpha_n, 1))$.

5.5 Computing b

The Seifert manifold $(F_g; \alpha_1,\ldots,\alpha_n)\backslash ST(H^2)$ is then

$$(Oog|-n-b; (\alpha_1, \alpha_1-1),\ldots,(\alpha_n, \alpha_n-1))$$

(compare 4.4). To obtain -n-b we use a method analogous to that of 4.6 (compare [RV]). There, we had that the *compact* SO(3) covers the manifolds of spherical tessellations. In the hyperbolic case, the natural covering $ST(H^2)$ is not compact, and we must search for an intermediate compact covering without exceptional fibers. Its existence is guaranteed by a result of Fox [Fo] (see also [LS], page 143), which asserts *the existence of a normal subgroup F' of* $(F_g; \alpha_1,\ldots,\alpha_n)$ *of finite index and such that* $F' = (F_h ; -)$, *i.e. F' does not contain elliptic transformations.*

Assuming the existence of F' we have the following commutative diagram

$$(F_h' ; -)\backslash ST(H^2) \rightarrow (F_g; \alpha_1,\ldots,\alpha_n)\backslash ST(H^2)$$
$$p' \downarrow \qquad\qquad\qquad p \downarrow$$
$$(F_h' ; -)\backslash H^2 \rightarrow (F_g; \alpha_1,\ldots,\alpha_n)\backslash H^2$$

where the horizontal maps are n-fold coverings and p, p' are Seifert fibered bundles.

As in 4.6 we have that $(F_h' ; -)\backslash ST(H^2)$, being the spherical tangent bundle of F_h' is $(Ooh|2h-2)$, where

$$2h-2 = k(-n-b + \frac{\alpha_1 - 1}{\alpha_1} + \ldots + \frac{\alpha_n - 1}{\alpha_n}) \ .$$

On the other hand, the area, $2\pi(2h-2)$, of $F'\backslash H^2$ is k times the area of $(F_g, \alpha_1, \ldots, \alpha_n)$, which is

$$-2\pi\chi(F_g; \alpha_1, \ldots, \alpha_n) = 2\pi(2g-2 + n - (\frac{1}{\alpha_1} + \ldots + \frac{1}{\alpha_n})) \ .$$

Hence b = 2-2g-n.

Thus

$$M(F_g; \alpha_1, \ldots, \alpha_n) = (Oog \mid 2-2g-n \ ; \ (\alpha_1, 1), \ldots, (\alpha_n, 1)) \ .$$

Exercise 1. Compute b *for the manifolds of euclidean tessellations using this procedure.*

Hint.- Take the subgroup of translations $\Gamma' \leq \Gamma$.

5.6 Involutions

Exactly as in 2.8 and 3.15, if a particular tessellation has an orientation-reversing symmetry, the corresponding manifold admits the involution induced by reflection through a hyperbolic line of H^2. The quotient space is also the spherical tangent bundle of an orbifold.

For instance the tessellation S732 of Fig. 4 gives rise to a reflection-symmetric tessellation by forgetting the shading. The manifold M(S732) admits the involution which is induced by reflection in any hyperbolic line. The quotient manifold (spherical tangent bundle of D$\overline{\overline{7}}\overline{3}2$) is the orbifold of Fig. 8, where the isotropy groups of the points in the knot are cyclic of order two.

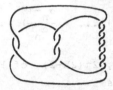

Fig. 8. ST(D$\overline{732}$)

Exercise 1. M(S732) *has an* S^1-*action (rotation around* O $\in$ H^2), *and* $\mathbb{Z}_2 \leq$ S^1 *defines an involution whose quotient space is* S^3. *Is this involution the same as the one induced by reflection in a line?*

Hint.- Find the singular set of the quotient orbifold, using the same idea as in Exercise 3 of 4.6.

Exercise 2. *The same as before for* S642 *(Fig. 5).*

Exercise 3. *Find the* $\mathbb{Z}_m$'s *embedded in* S^1 *which induce free actions on* M(F$_g$; $\alpha_1,\ldots,\alpha_n$). *Find the quotient manifolds.*

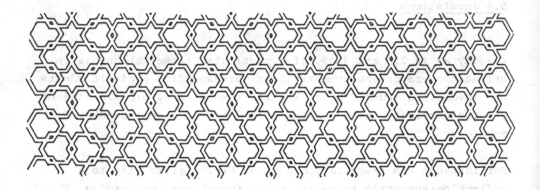

Appendix B: The Hyperbolic Plane

"Como las cosas humanas no sean eternas, yendo
siempre en declinación en sus principios, hasta
llegar a su último fin,..., llegó su fin y
acabamiento."

*"As all human things, especially the lives of men,
are transitory, being ever on the decline from
their beginnings till they reach their final end,
and as Don Quixote has no privilege from Heaven
exempting him from the common fate, his dissolution
and end came when he least expected it."*
Cervantes, Don Quixote, Part II, Ch. CXXVII, Of how
Don Quixote fell ill, of the Will he made, and of
his Death.

This Appendix should be read after acquiring some knowledge of
hyperbolic geometry (see, for instance, [VY], [Bl], [Le], [Cox2]).
The reader can also consult [Be]. We have included it because the
ideas it contains fit into the theme of the book and hopefully can
help to give a unified vision of the euclidean, elliptic and hyperbolic planes. This Appendix is not intended to be an introduction
to hyperbolic geometry, but should be considered as the first draft
for an alternative presentation of the hyperbolic plane. Thus it
consists of snapshots and ideas and it is necessarily incomplete.
The reader is advised, in addition, to consult [Th2] and [Sc] as
good complements. The sources for this appendix are [DuV] and [Mag].

B.1 The Klein model of H^2

Let Q be a quadric in $\mathbb{RP}^3$ of ellipsoidal type (think of a sphere
in $\mathbb{R}^3 \subset \mathbb{RP}^3$, for instance). Let O be a point in $\mathbb{RP}^3$ which we
imagine to be the eye of an observer. The set of lines passing
through O is a model of the projective plane $\mathbb{RP}^2$.

The relative position of O and Q gives rise to three different geometries. In the *elliptic plane* (or *Riemann sphere*), O lies *inside* Q, in the same way as a sky observer is surrounded by the vault of heaven. Following Klein the properties of $\mathbb{RP}^2$ (= lines through O) which remain invariant under the group of projective transformations of $\mathbb{RP}^3$ fixing O and Q (as sets), constitute the realm of *elliptic geometry*.

When O is *outside* Q we have the *hyperbolic* plane. The corresponding *group*, Iso(H^2), consists of projective transformations fixing Q and O. The lines passing through O (= points of $\mathbb{RP}^2$) decompose into three mutually disjoint classes: those cutting Q in two points (*points of H^2*); those tangent to Q (*points at infinity of H^2*); and those not cutting Q (*points at ultrainfinity of H^2*). The elements of Iso(H^2) leave the three sets invariant. It is customary to think of H^2 *as the set of lines through O cutting Q in two points, together with the group* Iso(H^2).

Instead of taking the rays through O we can think of H^2 as the intersections of these rays with the polar plane P of O with respect to Q (Fig. 1). This is the more familiar *Klein model of the hyperbolic plane*, namely the set of interior points of an ellipse of $\mathbb{RP}^2$ = P together with the group of projective transformations of $\mathbb{RP}^2$ fixing the ellipse. In fact, if a projective map fixes O and Q, then it also fixes P, hence P ∩ Q. Conversely, if a projective map fixes O, P and P ∩ Q, it fixes Q (it even fixes each element of the pencil of quadrics determined by Q and P taken twice).

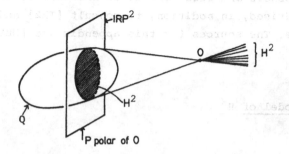

Fig. 1. H^2

If the point O lies *on* Q something beautiful happens. The group
of projective maps fixing Q and O, also fixes the polar P of O with
respect to Q, i.e. the tangent plane P of Q at O. The set of lines
passing through O, and outside that tangent plane P, have the
structure of an affine plane (the lines through O lying on P are
just the *points at infinity* of the affine plane), where the group of
affinities is the group of projective maps fixing P and O. The set
of affine maps also fixing Q forms a proper subgroup of the group
of affinities and defines in the affine plane $\mathbb{RP}^2$-P, a proper subgeometry of the affine geometry. We will show that it is the geometry of *similarities* (*parabolic* geometry) of the affine plane. Now
in *contrast* to the elliptic and hyperbolic cases, if a projective
map fixes P and Q, it does not need to fix the quadrics of the
pencil determined by Q and P taken twice (because Q and P^2 do not
define a reference in the pencil which is an $\mathbb{RP}^1$). Thus the subset of projective maps fixing *each* quadric of that pencil, form a
subgeometry of the similarity geometry of the affine plane. That
geometry, as we will show, is *euclidean geometry*.

Thus, euclidean geometry plays an intermediate role between hyperbolic and elliptic geometries. In passing from one to the other, the
pencil generated by Q and O degenerates when O is on Q. This creates
a "peculiar" degenerative property for euclidean geometry, namely,
the possibility of having infinitely many similar triangles which
are not congruent. In the other two geometries, the angles determine
the triangle.

The above considerations work perfectly well if we take $\mathbb{RP}^{n+1}$
instead of $\mathbb{RP}^3$. This gives the definition of the hyperbolic space
H^n, and of its group Iso(H^n) of transformations.

B.2 Pencils in H^2

Think of H^2 as the interior of a conic C in $\mathbb{RP}^2$. There are three
types of pencils: *elliptic*, *parabolic* or *hyperbolic*, according to
whether the vertex of the pencil is inside, on or outside the conic
(see Fig. 2). The *geometry of a pencil* is the one induced by the

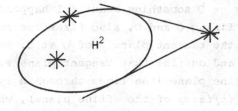

Fig. 2. Pencils of H^2

subgroup Iso(H^2m O) of Iso(H^2) fixing its vertex O. According to
the last section, *that geometry is the 1-dimensional elliptic, parabolic or hyperbolic geometry, respestively.* The lines of an elliptic,
parabolic or hyperbolic pencil are called *incident, parallel* or
ultraparallel.

Take an elliptic pencil with vertex O in the interior of C. Any
conic of the conic pencil determined by C and the polar P of O taken
twice, is left invariant by Iso(H^2, O). The conics of the pencil,
which are within C, are called *circles with center O* and their
geometry is called *spherical* (Fig. 3a).

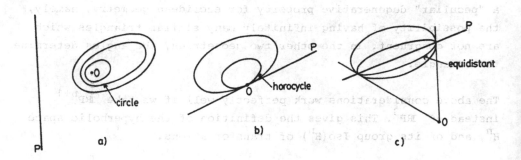

Figure 3

In the elliptic geometry defined by a point O within a hyperquadric

$$Q \equiv x_0^2 + x_1^2 + \cdots + x_n^2 - x_{n+1}^2 = 0$$

of $\mathbb{R}P^{n+1}$, a line and a hyperplane passing through O are called *orthogonal* if they are conjugate with respect to Q. Thus two lines in H^2 which meet at $O \in H^2$ are called *orthogonal* if they are orthogonal in the geometry of the pencil determined by O, i.e. if they are conjugate (with respect to C) (Fig. 4).

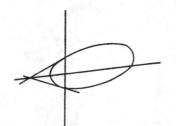

Fig. 4. Orthogonal lines

Take now a parabolic pencil with vertex at the point at infinity O (Fig. 3b). Any conic of the pencil of conics defined by the tangent P (taken twice) and C is called a *horocycle* if it lies within C. The geometry of the pencil of vertex O, given by $Iso(H^2, O)$, is parabolic, but the geometry of any horocycle D (given by $Iso(H^2, D)$) is *euclidean* (see section B.4).

Now consider a hyperbolic pencil with vertex in the point of ultra-infinity O (Fig. 3c). The polar P of O defines a line in H^2 (which has hyperbolic geometry). The hyperbolic lines which belong to the pencil, are orthogonal to P, because all of them are conjugated to P. Thus the *lines of the pencil all have a common perpendicular P* (compare Fig. 4, which shows that in the Klein model angles are distorted). Consider the pencil of conics determined by P (taken twice) and C. A conic D of that pencil lying within C has two points at infinity (common with P) and it is divided by them into two "curves" D_1, D_2 which are called *equidistants of P*. The geometry of D_i, given by

$$Iso(H^2, D) = Iso(H^2, O)$$

is hyperbolic.

Thus we see that H^2 contains spherical, euclidean and hyperbolic curves.

The terms "circle", "orthogonal", and "equidistant" suggest that the hyperbolic geometry of H^n is a *metric* geometry in the sense that there is a well-defined *distance* satisfying the usual conditions, or that H^n is a Riemannian manifold, such that the group of isometries is $Iso(H^n)$. This is the case, but for the moment we will just study properties of H^2 that can be obtained using the meager definition given before.

B.3 H^2 is strongly homogeneous

Given two pointers of H^2 (i.e. two elements of the spherical tangent bundle of H^2) there are precisely two elements of $Iso(H^2)$ sending one to the other. One element is orientation-preserving, the other reverses orientation. In fact, let v, v' be the two pointers based at a, a' with directions γ, γ' (Fig. 5). Let α, α' be the polar lines of a, a' and let b = α ∩ γ, b' = α' ∩ γ'. Finally, let β, β' be the polar lines of b, b', and let c = α ∩ β, c' = α' ∩ β'. The selfpolar triangle abc maps onto a'b'c' so that the segment s of γ containing v maps onto the segment s' of γ' containing v', i.e. d = s ∩ C maps onto d' = s' ∩ C. Now, there are two possibilities for a ∈ (ca) ∩ C, since it can be mapped onto either of the two points e', ê' of (c'a') ∩ C, v say e': then, there is precisely one projective map mapping (a b c d e) onto (a'b'c'd'e'). To see this it is enough to take as unit point of (abc) the point u such that uf and de separate harmonically (Fig. 5); then define u' analogously, and apply the von Staudt theorem to (a b c; u), (a'b'c'; u'). Thus the projective map defined above fixes C since it is subject to five conditions.

B.4 The elements of $Iso(H^2)$

An element φ of $Iso(H^2)$, being a projective map of RP^2, has at least one fixed point O. If O lies within C, then φ ∈ $Iso(H^2, O)$ and hence fixes the elliptic pencil with vertex O. Thus φ fixes the circles with center O and φ is called an *elliptic transformation* or *hyperbolic rotation* with center O. If, as is the case, H^2 has a

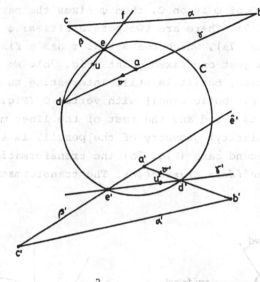

Fig. 5. H^2 is strongly homogenous

metric and φ is an isometry, the points of the circles with center O are the same distance from O, which explains the name "circle" (Fig. 6a).

If the fixed point of φ lies outside C, then φ fixes the hyperbolic pencil of vertex O; φ is called a *hyperbolic transformation* or *hyperbolic translation* because it fixes all the equidistant curves of P, the polar of O. As before, this explains the name "equidistant" for those curves (Fig. 6b).

a)

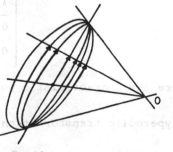

b)

Fig. 6. Transformations of H^2

If the fixed point O of φ is on C, then φ fixes the parabolic pencil
with vertex O (Fig. 7). There are two possibilities: φ is at the same
time hyperbolic (Fig. 7a), which means that φ has a fixed point O'
outside C, or φ has just one fixed point (Fig. 7b). We have just
studied the first case, but it is still interesting to understand the
effect of φ on the parabolic pencil with vertex O (Fig. 7a). The
line E polar to O' is fixed and the rest of the lines move. The
effect on the (similarity) geometry of the pencil, is a *homothety of
center E*. In the second case (Fig. 7b) the transformation induced on
the pencil is an *euclidean translation*. The transformation is called
parabolic.

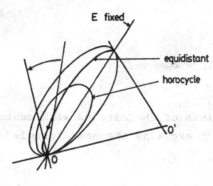

E fixed

equidistant

horocycle

a)   b)

Figure 7

Thus we see that a hyperbolic rotation has Jordan form

$$\begin{bmatrix} \lambda & 0 & 0 \\ 0 & & \\ 0 & & A \end{bmatrix}$$

where the characteristic polynomial of A is irreducible over $\mathbb{R}$.

A hyperbolic transformation has Jordan form

$$\begin{bmatrix} \lambda_1 & & \\ & \lambda_2 & \\ & & \lambda_3 \end{bmatrix}, \lambda_1 \neq \lambda_2, \lambda_1 \neq \lambda_3$$

where λ_2 might coincide with λ_3, in which case the axis is pointwise invariant. In this case the matrix is

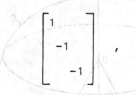

$$\begin{bmatrix} 1 & & \\ & -1 & \\ & & -1 \end{bmatrix},$$

the homology is harmonic (Fig. 8), and we have a *reflection through* **E**. If $\lambda_2 \neq \lambda_3$ we have the *hyperbolic translations* in the "strict sense".

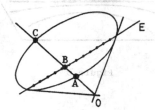

Fig. 8. OB separates AC harmonically

A parabolic transformation has Jordan form

$$\begin{bmatrix} \lambda & 1 & 0 \\ 0 & \lambda & 1 \\ 0 & 0 & \lambda \end{bmatrix}$$

Exercise 1. *Describe the transformations of* $\mathrm{Iso}(H^3)$.

Hint.- If the center O is outside C, the transformation induced on the H^2, polar of O with respect to C, is one of those studied above. If the center is on C we have an interesting case (Fig. 9). In this case, if the axis A is pointwise fixed, the transformation induced on the horosphere E is a rotation. If A is only invariant, the transformation induced on the similarity geometry of the pencil with vertex O is a *snail* transformation (composition of rotation and homothety).

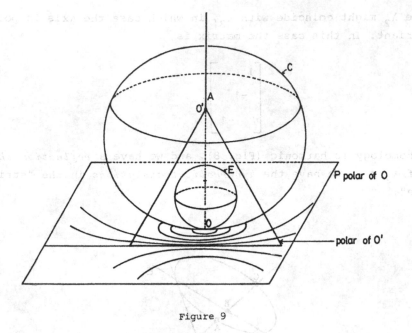

Figure 9

B.5 Metric

Let $(x:y:z:u)$ be homogeneous coordinates for $\mathbb{RP}^3$. Let C be given by the equation

$$x^2 + y^2 + z^2 - u^2 = 0 \ .$$

Assume, first, that O is within C. We can take O = (0:0:0:1). The polar of O is the plane P of equation u = 0. If we take affine coordinates

$$(X_1, X_2, X_3) = (\frac{x}{u}, \frac{y}{u}, \frac{z}{u}) \ ,$$

u = 0 becomes the plane at infinity, O is the origin and C is the sphere of radius 1 and center O. Moreover, the vector space $\mathbb{R}^3$ has the quadratic form

(1)
$$X_1^2 + X_2^2 + X_3^2$$

which defines a Riemannian structure on $\mathbb{R}^3$ as follows. Given a point A in $\mathbb{R}^3$, the tangent space at A can be identified with $\mathbb{R}^3$, and we can associate to that tangent space the inner product defined by (1), i.e.

$$x_1 x_1' + x_2 x_2' + x_3 x_3'$$

where (x_1, x_2, x_3), (x_1', x_2', x_3') are coordinates of two tangent vectors at A. This Riemannian structure on $\mathbb{R}^3$ is obviously the *euclidean structure* and we give C the *induced Riemannian structure*. Thus the geodesics in C are the great circles and the distance between points is the angle (in radians) measured on the geodesic through them, etc. ... The projective maps fixing O and C, also fix P, hence they are affine maps fixing O and C. Thus the elliptic group coincides with O(3), the group of isometries of the Riemannian manifold just defined: the elliptic geometry is thus a metric geometry.

The same happens with the hyperbolic plane H^2. Now O is outside C. Thus we take O = (0:0:1:0) so that its polar plane P has equation z = 0. Since we want to give a metric to C-P, we assume P is the plane at infinity, by taking affine coordinates

$$(X_1, X_2, X_3) = (\frac{x}{z}, \frac{y}{z}, \frac{u}{z}) \quad .$$

Thus O becomes the origin of $\mathbb{R}^3$ and C-P is the two-sheeted hyperboloid given by

$$x_1^2 + x_2^2 - x_3^2 = -1$$

and depicted in Fig. 10.

As before, the vector space $\mathbb{R}^3$ has the quadratic form $x_1^2 + x_2^2 - x_3^2$, which defines in $\mathbb{R}^3$ a pseudo-Riemannian structure (Lorentz's metric) because the bilinear form

(2) $$X_1 X_1' + X_2 X_2' - X_3 X_3'$$

is symmetric and nondegenerate (though it is not always positive). However C with the induced metric is a Riemannian manifold because the tangent plane at a point A ∈ C (see Fig. 10) can be identified with a vector plane of $\mathbb{R}^3$ lying outside the asymptotic cone of

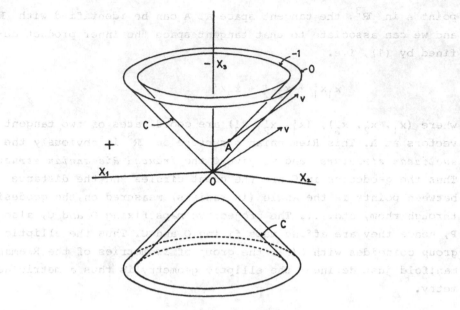

Fig. 10. The hyperbolic plane

C, where the bilinear form (2) is positive. Either sheet of C is a
model of H^2 and we have given it a Riemannian structure.

To convince ourselves that H^2 is metric, it only remains to show that
the group of isometries of H^2 is $\text{Iso}(H^2)$. Clearly, any element of
$\text{Iso}(H^2)$, being a projective map fixing O and C, defines an affine
map of $\mathbb{R}^3$ fixing O and C-P. Hence it leaves the form (2) invariant.
Thus it defines an isometry of the Riemannian manifold C. The converse is easier to prove using the contents of the next section.

To obtain the Riemannian metric on the "conic" model of H^2 we only
need to project C onto the plane $X_3 = 1$ from O (Fig. 11). Since we
only want to obtain a formula for the distance between two points of
H^2, we can work with the hyperbolic line (ignore the coordinate X_1 in
Fig. 10). Let Q be the point (0, 1). Defining t by

$$(X_2,\ X_3) = (t,\ 1)\quad ,$$

the projection of $X_3 = 1$ onto C from O is parametrized by

$$t \mapsto (t(1-t^2)^{-1/2},\ (1-t^2)^{-1/2})\ .$$

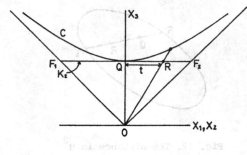

Fig. 11. The hyperbolic line

The differential of this map at the point t is:

$$dt \mapsto ((1-t^2)^{-3/2}, \; t(1-t^2)^{-3/2})dt \quad .$$

Hence the quadratic form $x_2^2 - x_3^2$ defines the form

$$((1-t^2)^{-3} - t^2(1-t^2)^{-3}) \; dt^2 = \frac{dt^2}{(1-t^2)^2}$$

on the tangent plane at $t \in (X_3 = 1)$. Thus the distance between $Q = (0, 1)$ and $R = (t, 1)$ is

$$d(Q, R) = \int_0^t \frac{dt}{(1-t^2)} = \frac{1}{2} \log \frac{1+t}{1-t} \quad .$$

If the cross ratio of four points (A, B, C, D) on a line is defined by

$$\frac{A - B}{B - C} \cdot \frac{C - D}{D - A}$$

we have

$$d(Q, R) = \frac{1}{2} \left| \log (Q, F_2, R, F_1) \right|,$$

where F_2, F_1 are the points at which $\overline{QR}$ meets C (Fig. 12).

Given two arbitrary points of H^2, Q' and R', we have seen in B.3 that there is an element φ of $\mathrm{Iso}(H^2)$ sending Q' to Q and R' to some

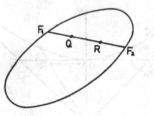

Fig. 12. The distance in H^2

R on the line in which we have made the above calculation. The
points F_1', F_2' at which $\overline{Q'R'}$ meets C are mapped onto F_1, F_2. Since
φ is an isometry of H^2,

$$d(Q',R')=d(Q,\ R)=\frac{1}{2}\ |\log\ (Q,\ F_2,\ R,\ F_1)|=\frac{1}{2}|\log\ (Q',\ F_2',\ R',\ F_1')|$$

because the cross ratio is a projective invariant. Hence *the distance*
$d(Q,\ R)$ *in* H^2 *is given by*

$$\frac{1}{2}\ |\log\ (Q,\ F_1,\ R,\ F_2)|\qquad.$$

B.6 The complex projective line

The purpose of this and the next section is to show that the group
of projective maps of $\mathbb{R}P^3$ fixing the quadric C, and the group of
homographies of the *complex projective line* $\mathbb{C}P^1$ are closely related.
This will allow us to obtain a new model of H^2 based on complex numbers (the Poincaré model in which angles are not distorted).

The *complex projective line* $\mathbb{C}P^1$ is the set of complex lines passing
through $(0,\ 0)$ in $\mathbb{C}\times\mathbb{C}$ with the quotient topology. We identify $\mathbb{C}P^1$
with the one-point compactification of $\mathbb{C}$ by the maps

$$\mathbb{C}+\infty\rightarrow\mathbb{C}P^1$$

$$z\mapsto(z:1)$$

$$\infty\mapsto(1:0)$$

$$\mathbb{C}P^1 \to \mathbb{C} + \infty$$

$$(z_1 : z_2) \mapsto z_1/z_2 \qquad (z_1/0 := \infty)$$

The *cross-ratio* of four points (z_1, z_2, z_3, z_4) of $\mathbb{C}P^1$ is defined by

$$(z_1, z_2, z_3, z_4) = \frac{z_1 - z_2}{z_2 - z_3} \frac{z_3 - z_4}{z_4 - z_1}$$

whose meaning is better grasped by noticing that the argument of (z_1, z_2, z_3, z_4) is zero (i.e. (z_1, z_2, z_3, z_4) is real) if and only if z_1, z_2, z_3, z_4 lie on a circle in $\mathbb{C} + \infty$ (where "circle" also means straight line = circle of "infinite" radius). This is an exercise for the reader (Fig. 13).

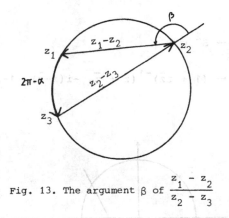

Fig. 13. The argument β of $\dfrac{z_1 - z_2}{z_2 - z_3}$

We are interested only in those projective maps of $\mathbb{C}P^1$ which induce on the base field $\mathbb{C}$ the identity automorphism or the *conjugation automorphism* sending z to $\bar{z}$. The first type of projective maps will be called *homographies* and the second type will be referred to as *antihomographies*. Once their equations are normalized they look like:

$$z \mapsto z' = \frac{\alpha z + \beta}{\gamma z + \delta} \quad ,$$

$$z \mapsto z' = \frac{\alpha \bar{z} + \beta}{\gamma \bar{z} + \delta}$$

respectively, where α, β, γ, $\delta \in \mathbb{C}$ and $\alpha\delta - \beta\gamma = 1$.

The set of all these projective maps forms a group G*, and since $z \mapsto \bar{z}$ is an antihomography, we see that the set of homographies G is a subgroup of G* of index two.

The homographies preserve the cross-ratio; the antihomographies conjugate it. In both cases it remains invariant when it is real. Thus *the elements of G* transform circles into circles.*

B.7 The stereographic projection

Let S^2 be the unit sphere in $\mathbb{R}^3$, and identify $\mathbb{R}^2 \times 0$ with $\mathbb{C}$ by sending $(a, b, 0)$ to $a + bi$. The projection of S^2 onto $\mathbb{R}^2 \times 0$ from the south pole $(0, 0, -1)$ is the *stereographic projection*. It defines the map (Fig. 14):

$$\mathbb{C}P^1 \to S^2 \subset \mathbb{R}^3$$

$$z \mapsto (1 + z\bar{z})^{-1} (z + \bar{z}, -i(z-\bar{z}), 1-z\bar{z})$$

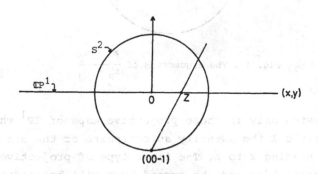

Fig. 14. Stereographic projection

In the next section we want to show that *the stereographic projection defines an isomorphism between the group G* and the group of projective maps of* $\mathbb{R}P^3$ *fixing* S^2.

To do this, we will give a more conceptual definition of the stereo-graphic projection. The sphere S^2 has the equation

$$x^2 + y^2 + z^2 = u^2$$

using homogeneous coordinates $(x:y:z:u)$ for $\mathbb{R}P^3$. In $\mathbb{C}P^3$ this equation represents a ruled quadric A which can be written

$$(x + iy)(x - iy) = (u + z)(u - z)$$

so that its two pencils of lines are given by (Fig. 15)

(1)
$$\left\{ \begin{array}{l} \lambda_2(x + iy) = (u + z)\lambda_1 \\ \\ \lambda_1(x - iy) = (u - z)\lambda_2 \ , \end{array} \right. \quad (\lambda_1 : \lambda_2) \in \mathbb{C}P^1$$
(2)

and

(3)
$$\left\{ \begin{array}{l} \mu_1(x + iy) = (u - z)\mu_2 \\ \\ \mu_2(x - iy) = (u + z)\mu_1 \ , \end{array} \right. \quad (\mu_1 : \mu_2) \in \mathbb{C}P^1 .$$
(4)

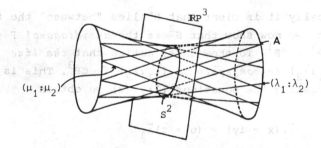

Fig. 15. The quadric A

The quadric A is topologically $\mathbb{C}P^1 \times \mathbb{C}P^1$ and, in fact, the map

$$f: \mathbb{C}P^1 \times \mathbb{C}P^1 \to A$$

$$[(\lambda_1 : \lambda_2), (\mu_1 : \mu_2)] \mapsto (\lambda_1\mu_2 + \lambda_2\mu_1 : -i(\lambda_1\mu_2 - \lambda_2\mu_1) : \lambda_2\mu_2 - \lambda_1\mu_1 : \lambda_2\mu_2 + \lambda_1\mu_1)$$

is a homeomorphism, since from (1) and (4) one gets

$$x = \frac{1}{2} (u + z)(\lambda_1/\lambda_2 + \mu_1/\mu_2)$$

$$iy = \frac{1}{2} (u + z)(\lambda_1/\lambda_2 - \mu_1/\mu_2)$$

and from (2) and (3)

$$x = \frac{1}{2} (u - z)(\lambda_2/\lambda_1 + \mu_2/\mu_1)$$

$$iy = \frac{1}{2} (u - z)(\lambda_2/\lambda_1 - \mu_2/\mu_1) \quad .$$

From this we obtain

$$(u + z)(\lambda_1/\lambda_2 + \mu_1/\mu_2) = (u - z)(\lambda_2/\lambda_1 + \mu_2/\mu_1) \quad .$$

Hence

$$(u + z)/(u - z) = \lambda_2 \mu_2/\lambda_1 \mu_1 \quad .$$

And it is an easy exercise to obtain the equation of f.

Now geometrically it is clear that S^2 lies "between" the two pencils of A. In fact, we now show that S^2 *is the antidiagonal* $\Gamma = \{(\lambda_1:\lambda_2),$ $(\overline{\lambda}_1:\overline{\lambda}_2)\} \subset \mathbb{C}P^1 \times \mathbb{C}P^1$. To show this, notice that *the line* $(\mu_1:\mu_2) = (\overline{\lambda}_1:\overline{\lambda}_2)$ *is conjugate to* $(\lambda_1:\lambda_2)$ *in* $\mathbb{C}P^3$. This is obvious, because conjugating the equations (1), (2) we obtain

$$\overline{\lambda}_2 (x - iy) = (u + z)\overline{\lambda}_1$$

$$\overline{\lambda}_1 (x + iy) = (u - z)\overline{\lambda}_2$$

which define the line $(\mu_1:\mu_2) = (\overline{\lambda}_1:\overline{\lambda}_2)$, using the equations (3), (4). Thus f maps Γ onto S^2, because two conjugate lines cut each other in a real point.

We have proved that the map

$$abf : \mathbb{C}P^1 \xrightarrow{} \Gamma \xrightarrow{f} S^2$$

$$(\lambda_1:\lambda_2) \mapsto [(\lambda_1:\lambda_2),\ (\overline{\lambda}_1:\overline{\lambda}_2)] \to f[(\lambda_1:\lambda_2),\ (\overline{\lambda}_1:\overline{\lambda}_2)]$$

is a homeomorphism.

By letting $z = (\lambda_1:\lambda_2)$, the map abf is the stereographic projection given at the beginning of this section. This is our promised inter-pretation of stereographic projection.

B.8 Interpreting G*

The elements of G* act on $\mathbb{C}P^1 = \Gamma$. A homography acts as follows:

$$\left(\begin{bmatrix} \lambda_1 \\ \lambda_2 \end{bmatrix},\ \begin{bmatrix} \overline{\lambda}_1 \\ \overline{\lambda}_2 \end{bmatrix} \right) \to \left(\begin{bmatrix} \alpha & \beta \\ \gamma & \delta \end{bmatrix} \begin{bmatrix} \lambda_1 \\ \lambda_2 \end{bmatrix},\ \begin{bmatrix} \overline{\alpha} & \overline{\beta} \\ \overline{\gamma} & \overline{\delta} \end{bmatrix} \begin{bmatrix} \overline{\lambda}_1 \\ \overline{\lambda}_2 \end{bmatrix} \right)$$

while an antihomography acts as follows

$$\left(\begin{bmatrix} \lambda_1 \\ \lambda_2 \end{bmatrix},\ \begin{bmatrix} \overline{\lambda}_1 \\ \overline{\lambda}_2 \end{bmatrix} \right) \to \left(\begin{bmatrix} \alpha & \beta \\ \gamma & \delta \end{bmatrix} \begin{bmatrix} \overline{\lambda}_1 \\ \overline{\lambda}_2 \end{bmatrix},\ \begin{bmatrix} \overline{\alpha} & \overline{\beta} \\ \overline{\gamma} & \overline{\delta} \end{bmatrix} \begin{bmatrix} \lambda_1 \\ \lambda_2 \end{bmatrix} \right).$$

These homographies and antihomographies of Γ extend in a unique way, to transformations of $(\mathbb{C}P^1 \times \mathbb{C}P^1,\ \Gamma)$ by replacing $\begin{bmatrix} \overline{\lambda}_1 \\ \overline{\lambda}_2 \end{bmatrix}$ by $\begin{bmatrix} \mu_1 \\ \mu_2 \end{bmatrix}$ in the previous equations. These transformations of $\mathbb{C}P^1 \times \mathbb{C}P^1$ map $\lambda_1\mu_1,\ \lambda_2\mu_1,\ \lambda_1\mu_2,\ \lambda_2\mu_2$ into linear combinations of themselves. Thus, for instance:

$$\lambda_1\mu_1 \mapsto (\alpha\lambda_1 + \beta\lambda_2)(\overline{\alpha}\mu_1 + \overline{\beta}\mu_2) = \alpha\overline{\alpha}\lambda_1\mu_1 + \beta\overline{\beta}\lambda_2\mu_2 + \alpha\overline{\beta}\lambda_1\mu_2 + \beta\overline{\alpha}\lambda_2\mu_1 \ .$$

Hence, these transformations of $(\mathbb{C}P^1 \times \mathbb{C}P^1,\ \Gamma)$ induce, via

$$f: (\mathbb{C}P^1 \times \mathbb{C}P^1,\ \Gamma) \to (A,\ S^2)\ ,$$

transformations of $(A,\ S^2)$ which extend to homographies of $(\mathbb{C}P^3,\ \mathbb{R}P^3)$. These extensions are unique because S^2 contains a projective reference of $\mathbb{R}P^3$. We finally remark that the extensions coming from a homography (resp. antihomography) of $\mathbb{C}P^1$ preserve (resp. permute) the two pencils of A.

Conversely, let φ be a homography of $(\mathbb{C}P^3, \mathbb{R}P^3, S^2)$. Then φ fixes A and it preserves or permutes the two pencils of A. Thus φ induces a map on the pencils:

$$\lambda \mapsto \frac{a\lambda + b}{c\lambda + d} \quad , \quad \mu \mapsto \frac{p\mu + q}{r\mu + s}$$

if it preserves the pencils, or

$$\lambda \mapsto \frac{a\mu + b}{c\mu + d} \quad , \quad \mu \mapsto \frac{p\lambda + q}{r\lambda + s}$$

otherwise. Since φ fixes $S^2 = \{(\lambda, \bar{\lambda})\}$ then, in the first case the second coordinate of

$$\left(\frac{a\lambda + b}{c\lambda + d} \quad , \quad \frac{p\bar{\lambda} + q}{r\bar{\lambda} + s} \right)$$

is conjugate to the first coordinate. Thus, in the first case $a = \bar{p}$, $b = \bar{q}$, $c = \bar{r}$, $d = \bar{s}$. Hence φ defines a homography of Γ. In the second case, it is easy to see that φ defines an antihomography of Γ.

We have proved that *the group G^* consists of the homographies of $\mathbb{R}P^3$ fixing S^2. Hence G^* can be identified with the group Iso (H^3). More-over the relation between both groups is established via the stereo-graphic projection of $\mathbb{C}P^1$ onto S^2.*

B.9 Digression: types of isometries of H^3

The above result is very important. With it, we can understand readily the types of elements of Iso (H^3) just by studying the elements of

G*. This is what we will do in this section. In the next two
sections we will study the subgroups of G* which under stereographic
projection fix a point $O \in \mathbb{RP}^3$. This will allow us to interpret the
elliptic, parabolic and hyperbolic groups in two dimensions as subgroups of G*, and to obtain models of the elliptic, parabolic and
hyperbolic planes based on complex numbers.

Let us start studying the elements of G*. We say that a line of
$\mathbb{CP}^3$ has *real origin* if it is a complexified line of $\mathbb{RP}^3$.

The double points of $\lambda' = \dfrac{\alpha\lambda + \beta}{\gamma\lambda + \delta}$ are the roots λ_1, λ_2 of

$$\gamma\lambda^2 + (\delta - \alpha)\lambda - \beta = 0 .$$

Then, the lines $\lambda = \lambda_1$, λ_2 and $\mu = \bar{\lambda}_1$, $\bar{\lambda}_2$ of A remain setwise fixed
and there are 4 double points, two real $(\lambda_1, \bar{\lambda}_1)$, $(\lambda_2, \bar{\lambda}_2)$, and two
others, $(\lambda_1, \bar{\lambda}_2)$ and $(\lambda_2, \bar{\lambda}_1)$, which are conjugate (Fig. 16). They
determine two double lines of real origin, namely

$$\pi = (\lambda_1, \bar{\lambda}_1) \cup (\lambda_2, \bar{\lambda}_2)$$

and

$$\pi' = (\lambda_1, \bar{\lambda}_2) \cup (\lambda_2, \bar{\lambda}_1) .$$

Since (polar π) = polar $(\lambda_1, \bar{\lambda}_1)$ ∩ polar $(\lambda_2, \bar{\lambda}_2)$ = {tangent plane
to A at $(\lambda_1, \bar{\lambda}_1)$} ∩ {tangent plane to A at $(\lambda_2, \bar{\lambda}_2)$} = π' we deduce
that

$$e := \pi \cap \mathbb{RP}^3$$

and

$$e' := \pi' \cap \mathbb{RP}^3$$

are polar to each other with respect to S^2, and e contains the two
double real points of φ, which lie on S^2. We say that e *is the
primary axis and* e' *the secondary axis* of the homography (Fig. 17).

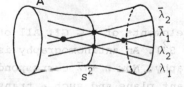

Fig. 16. The double points of a homography

206

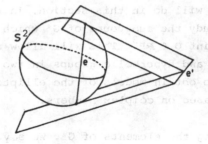

Fig. 17. Primary and secondary axis

The possible cases are diagramatically depicted in Figs. 18 - 20
(compare Fig. 9 and Fig. 18). The only one requiring explanation is
the parabolic case. Here the double points coincide and if we
assume that that double point is the south pole, then under stereographic projection we obtain a homography of $\mathbb{C}P^1$ fixing ∞. That map
is of the form $z \mapsto z + \beta$ i.e. a *translation*. Hence there is a
pencil of invariant planes passing through an invariant line L
(Fig. 20a), which cannot be pointwise fixed (otherwise there would
be invariant lines outside P). This implies the existence of a pointwise fixed line on P. These lines are the *primary* and *secondary* axes.

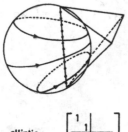

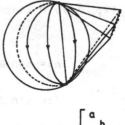

elliptic : $\begin{bmatrix} 1 & & & \\ & 1 & & \\ & & 0 & 1 \\ & & -1 & 0 \end{bmatrix}$

hyperbolic : $\begin{bmatrix} a & & & \\ & b & & \\ & & 1 & \\ & & & 1 \end{bmatrix}$

Figure 18

We have described the elements of $G \subset G^*$. All of them preserve the
orientation of $\mathbb{C}P^1$ (or S^2). An antihomography is the composition of
$z \mapsto \bar{z}$ with some homography. Now $z \to \bar{z}$ corresponds to an involution
in $\mathbb{R}P^3$ with an invariant plane and such a transformation is

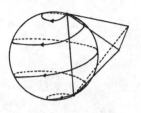

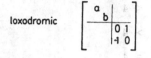

loxodromic $\begin{bmatrix} a \\ \ b \\ \hline & 0 \ 1 \\ & -1 \ 0 \end{bmatrix}$

involutive $\begin{bmatrix} a \\ \ \ a \\ \ \ \ \ \ b \\ \ \ \ \ \ \ \ b \end{bmatrix}$

Figure 19

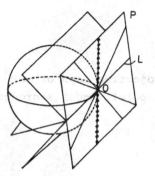

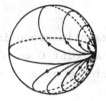

a) parabolic $\begin{bmatrix} a & 1 \\ & a & 1 \\ & & a & 1 \\ & & & a \end{bmatrix}$ b)

Figure 20

orientation-reversing in $\mathbb{RP}^3$, S^2 and $\mathbb{CP}^1$. Thus G *is the subgroup of orientation-preserving transformations of* G*.

Exercise 1. Every homography of $\mathbb{CP}^1$ *is conjugate to one of the following*

$$z' \mapsto kz \begin{cases} k = e^{ia} & elliptic \\ k\ real > 0 & hyperbolic \\ k = re^{ia} & loxodromic \\ k = -1 & involutive \end{cases}$$

$$z' \mapsto z + b \qquad parabolic$$

Moreover

$$z' \mapsto \frac{\alpha z + \beta}{\gamma z + \delta} \ , \ \alpha\delta - \beta\gamma = 1$$

is

elliptic if and only if $(\alpha + \delta)^2 < 4$

hyperbolic   " $(\alpha + \delta)^2 > 4$

parabolic   " $(\alpha + \delta)^2 = 4$

loxodromic   " $(\alpha + \delta)^2$ *is not real* .

B.10 The parabolic group

The parabolic group is the group of projective maps of $\mathbb{R}P^3$ fixing S^2 and a point O on S^2 (e.g. the south pole). Under stereographic projection O goes to ∞ in

$$\mathbb{C}P^1 = \mathbb{C} + \infty \ \ .$$

The parabolic group is thus the subgroup of G^* *fixing* ∞, and the *parabolic geometry is the geometry of* $\mathbb{C}$ under the action of the parabolic group. Now an element of G fixes ∞ if and only if it is of the form $z \mapsto \alpha z + \beta$. This is the *similarity group* of $\mathbb{C} \cong \mathbb{R}^2$ as we predicted in B.1.

The group fixing S^2, the south pole O, the quadrics of the pencil determined by S^2 and the tangent plane of S^2 at O, is composed of parabolic and elliptic transformations, because the primary axis must be pointwise fixed. But these transformations go, under stereographic projection, to *translations* and *rotations* of $\mathbb{C}$. Thus, *the geometry of a horosphere is the geometry of the euclidean plane.*

B.11 The elliptic group

The elliptic group is the group of projective maps of $\mathbb{R}P^3$ fixing S^2 and some point O inside S^2; or, equivalently, fixing S^2 and commuting with the harmonic homology with center O and having as invariant plane the polar of O. Taking O as the center of S^2, the principal axis must be pointwise fixed, and we see that the elliptic group is O(3).

The condition of commuting with the harmonic homology with center O and having as invariant plane the plane at infinity (i.e. the map $z \mapsto -\frac{1}{\bar{z}}$ in $\mathbb{CP}^1$) implies that

$$z' = \frac{\alpha z + \beta}{\gamma z + \delta}$$

has the form

$$z' = \frac{\alpha z + \beta}{-\bar{\beta} z + \bar{\alpha}} \quad , \quad \alpha\bar{\alpha} + \beta\bar{\beta} = 1$$

Now, the matrices

$$\begin{bmatrix} \alpha & \beta \\ -\bar{\beta} & \bar{\alpha} \end{bmatrix}$$

with $\alpha\bar{\alpha} + \beta\bar{\beta} = 1$, define the group SU(2) and we have the 2-fold covering

$$SU(2) \to SO(3)$$

$$\begin{bmatrix} \alpha & \beta \\ -\bar{\beta} & \bar{\alpha} \end{bmatrix} \mapsto z' = \frac{\alpha z + \beta}{-\bar{\beta} z + \bar{\alpha}}$$

Thus *the orientation-preserving maps of the elliptic group form a group isomorphic to* SO(3) = SU(2)/(+1, -1).

Exercise 1. Show that the map

$$\varphi: S^3 \to 1 \times S^3 \xrightarrow{\lambda} SO(4)$$

$$e^{\alpha P} \mapsto (1, e^{\alpha P}) \mapsto \{x \mapsto xe^{-\alpha P}\}$$

where λ is as in 3.6, is a homeomorphism onto the image. Show that the image is SU(2).

Hint.- Identify $\mathbb{C}^2$ with $\mathbb{Q}$ by $(a, b) \equiv (a + bj)$, and show that

$$\varphi(a + bj) = \begin{bmatrix} \bar{a} & \bar{b} \\ -b & a \end{bmatrix} .$$

Remark. This exercise shows that the map $\varphi(e^{\alpha P})$, which is a left-helix turn, preserves fibers of the Hopf fibration, hence it defines a map of $\mathbb{C}P^1$ (which is a rotation). The fixed points of this rotation correspond to two common fibers of the Hopf fibration and the left Hopf-like fibration induced by $\varphi(e^{\alpha P})$ (with axis P) (Fig. 21).

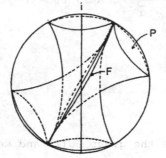

Fig. 21. Constructing a common fiber

B. 12 The hyperbolic group

The hyperbolic group, Iso (H^2), is the group of projective maps of $\mathbb{R}P^3$ fixing S^2 and a point O outside S^2, or equivalently, fixing S^2 and commuting with the harmonic homology with center O having P, the polar of O, as invariant plane. Since P cuts S^2 in a circle, this homology projects into an involution of $\mathbb{C}P^1$ fixing a line.

Taking this involution to be $z \mapsto \bar{z}$, the orientation-preserving subgroup Iso$^+$(H^2) is then given by the homographies

$$z' = \frac{az + b}{cz + d} \qquad a, b, c, d \text{ real}, \qquad ad - bc = 1 .$$

Thus Iso$^+$(H^2) is PSL(2, $\mathbb{R}$). As a model for H^2 one can take the half-plane (*half-plane model*)

$$E_2 = \{(x + iy) \in \mathbb{C} \,|\, y > 0\} \quad .$$

If instead of the involution $z \mapsto \bar{z}$ one takes its conjugate $z \to \dfrac{1}{z}$,

the group $\mathrm{Iso}^+(H^2)$ corresponds to

$$z' = \frac{\alpha z + \beta}{\bar{\beta} z + \bar{\alpha}} \quad , \quad \alpha\bar{\alpha} - \beta\bar{\beta} = 1 \quad .$$

Thus $\mathrm{Iso}^+(H^2)$ is $\mathrm{SU}(1, 1)/(+1, -1)$. As model for H^2 one can take the interior of the unit circle (*Poincaré model*)

$$P_2 = \{re^{i\theta} \in \mathbb{C} \,|\, r < 1\} \quad .$$

The transition from the half-plane model to the Klein model, and from this to the Poincaré model is given in Figures 22-25.

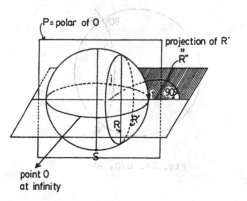

Fig. 22. Straight line in the half-plane model

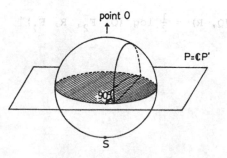

Fig. 23. Straight line in the Poincaré model

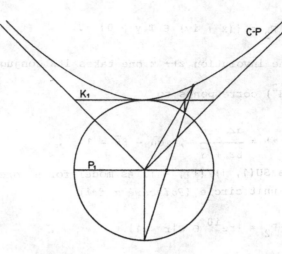

Fig. 24. The three models

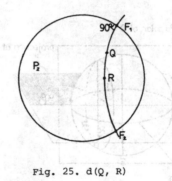

Fig. 25. d(Q, R)

Exactly the same argument as in B.5 shows that the distance between Q and R in the Poincaré model is given by the formula

$$d(Q, R) = \frac{1}{2} |\log (Q, F_2, R, F_1)|$$

where the cross-ratio is computed in $\mathbb{C}P^1$, and F_1, F_2 are as in Fig. 25.

The quadratic form giving the Riemannian structure of H^2 in the Poincaré model is computed as in B.5:

$$ds^2 = \frac{dx^2 + dy^2}{(1 - (x^2 + y^2))^2}$$

where the Poincaré model has real coordinates (x, y). We see that the metric is the euclidean one but with a magnification factor which is the distance of the point to the unit circle. The form of the metric immediately shows that the hyperbolic angle between two lines in P_2 is equal to the euclidean angle. This property is not shared by the Klein model where angles appeared distorted (remember Fig. 4). The fact that an isometry of the model P_2 (i.e. a differentiable map of P_2 preserving the metric) must preserve the magnitude of the angles (is conformal) implies that it is holomorphic or antiholomorphic, hence it must be a homography or anti-homography ([Ca]). We have thus proved that *the hyperbolic group coincides with the group of isometries with respect to the metric obtained above.*

Final remark. The study we have made is necessarily partial. We refer the reader to [Mi], [Th2] or [Sc] as a continuation. An important fact is that the hyperbolic plane has constant negative curvature, which is positive for the elliptic plane and zero for the euclidean plane. These results can be obtained by the reader as an exercise (see [DoC]). A consequence is that the sum of interior angles of a triangle is less than π. The difference from π (called the *defect*) is precisely the *area* of the triangle (Fig. 26). We see that the angles determine the triangle in the hyperbolic plane.

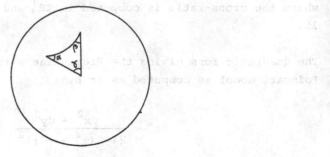

Fig. 26. Area = $\pi - (\alpha + \beta + \gamma)$

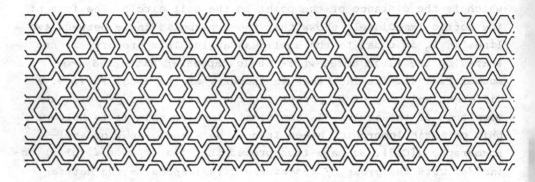

Source of the Ornaments Placed at the End of the Chapters

Preface. Mesones de Isuela (Zaragoza). Ceiling of the Castle's chapel

Chapter One. The "Seo" (Zaragoza). Wall in St. Michael's chapel

Chapter Two. The "Aljafería" (Zaragoza). Balustrade in the mosque

Appendix A. "Colegiata de Santa María la Mayor". Caspe (Zaragoza). Vault

Chapter Three. The "Alhambra" (Granada). Tiles in the "Patio de los Arrayanes".

Chapter Four. The "Seo" (Zaragoza). Door

Chapter Five. Tobed (Zaragoza). Church

Appendix B. Tamarite de Litera (Huesca). Door in the Church of St. Michael

References. Cervera de la Cañada (Zaragoza). Church

References

[Be]   BEARDON, A.F.: The geometry of discrete groups. Berlin-
          Heidelberg-New York: Springer 1983

[BS]   BONAHON, F., SIEBENMANN, L.: Seifert orbifolds and their
          role as natural crystalline parts of arbitrary compact
          irreducible 3-orbifolds. In: 1982 Sussex Conference
          proceedings, R. Fenn, ed. London Math. Soc. Lect. Notes
          series 95, pp. 18-85. Cambridge: Cambridge Univ. Press 1985

[Bl]   BLASCHKE, W.: Projektive Geometrie. Basel: Birkhäuser 1954

[BK]   BRIESKORN, E., KNÖRRER, H.: Plane algebraic curves. Basel-
          Boston-Stuttgart: Birkhäuser, 1986

[Ca]   CARTAN, H.: Théorie élémentaire des fonctions analytiques
          d'une ou plusieurs variables complexes. Paris: Hermann
          1978

[Cox1]   COXETER, H.S.M.: Regular Complex Polytopes. Cambridge:
          Cambridge Univ. Press 1974

[Cox2]   COXETER, H.S.M.: Introduction to geometry. New York: Wiley
          1969

[DoC]   DO CARMO, M.P.: Geometria Riemanniana. Rio de Janeiro:
          IMPA 1979

[Du1]   DUMBAR, W.: Fibered orbifolds and crystallographic groups.
          Ph. D. Thesis: Princeton 1981

[Du2]   DUMBAR, W.: Geometric orbifolds (preprint)

[DuV]   DU VAL, P.: Homographies, quaternions and rotations.
          Oxford: Clarendon Press 1964

[Fo] FOX, R.H.: On Fenchel's conjecture about F-groups. Mat. Tidskr. 8, 61-65 (1952)

[Fu] FULTON, W.: Algebraic curves. Reading, Mass.: Benjamin 1974

[GH] GREENBERG, M.J., HARPER, J.R.: Algebraic Topology; a first course. Reading, Mass.: Benjamin 1981

[GP] GUILLEMIN, V., POLLACK, A.: Differential Topology. Englewood Cliffs, N.J.: Prentice Hall 1974

[HC] HILBERT, D., COHN-VOSSEN, S.: Geometry and the imagination. New York-London: Academic Press 1952

[Jo] JOHNSON, D.L.: Topics in the theory of group presentations. Cambridge: Cambridge Univ. Press 1980

[Le] LEVY, H.: Projective and related geometries. Allendoerfer Advanced Series. New York: Macmillan 1964

[LS] LYNDON, R.C., SCHUPP, P.E.: Combinatorial group theory. Berlin-Heidelberg-New York: Springer 1977

[Mag] MAGNUS, W.: Non-euclidean tesselations and their groups. New York-London: Academic Press 1974

[Ma] MASSEY, W.S.: Algebraic Topology: an introduction. Berlin-Heidelberg-New York: Springer 1980

[Mi] MILNOR, J.: Hyperbolic geometry: The first 150 years. Bull. Amer. Math. Soc. 6, 9-23 (1982)

[Mo] MONTESINOS, J.: Variedades de Seifert que son recubridores cíclicos ramificados de dos hojas. Boletín Soc. Mat. Mex. 18, 1-32 (1973)

[RV] RAYMOND, F., VASQUEZ, A.T.: 3-manifolds whose universal coverings are Lie groups. Topology Appl.. 12, 161-179 (1981)

[R]   ROLFSEN, D.: Knots and links. Berkeley, Cal.: Publish or
            Perish 1976

[Rot]   ROTMAN, J.J.: The theory of groups. Boston, Mass.: Allyn
            and Bacon, Inc. 1973

[RS]   ROURKE, C.P., SANDERSON, B.J.: Introduction to Piecewise-
            linear-Topology. Berlin-Heidelberg-New York: Springer
            1982

[Sc]   SCOTT, P.: The geometries of 3-manifolds. Bull. London
            Math. Soc. 15, 401-487 (1983)

[S]   SEIFERT, H.: Topology of 3-dimensional fibered spaces
            (english translation of the german original in Acta Math.
            60, 147-288 (1933)). In [ST] pp. 359-422

[ST]   SEIFERT, H., THRELFALL, W.: A textbook of topology. New
            York-London: Academic Press 1980

[Sl]   SLODOWY, P.: Platonic solids, Kleinian singularities and
            Lie groups. In: Algebraic Geometry, Proceedings, Ann
            Arbor, 1981, Ed. I. Dolgachev, pp. 102-138, Lect. Notes
            in Math. 1008. Berlin-Heidelberg-New York: Springer 1983

[St]   STEENROD, N.: The topology of fiber bundles. Princeton,
            New Jersey: Princeton Univ. Press 1974

[TS]   THRELFALL, W., SEIFERT, H.: Topologische Untersuchung der
            Diskontinuitätsbereiche endlicher Bewegungsgruppen des
            dreidimensionalen sphärischen Raumes. Math. Ann. 104,
            1-70 (1930) and ibid., 107, 543-586 (1932)

[Th1]   THURSTON, W., WEEKS, J.R.: The mathematics of three-
            dimensional manifolds. Scientific American (July 1984)

[Th2]   THURSTON, W.: The geometry and topology of 3-manifolds.
            Princeton, New Jersey: Princeton Univ. Press (to appear)

[VY]   VEBLEN, O., YOUNG, J.W.: Projective geometry. I, II,
            Boston, Mass.: Ginn, 1910, 1918

[Wa]   WALDHAUSEN, F.: Eine Klasse von 3-dimensionalen Mannig-
faltigkeiten. I, II Invent. Math. 3, 308-333 (1967) and
ibid., 4, 501-504 (1967)

[Z]   ZIESCHANG, H.: Finite groups of mapping classes of
surfaces. Lect. Notes in Math. 875, Berlin-Heidelberg-
New York: Springer 1981

Further Reading

Books on low dimensional topology:

BIRMAN, J.S.: Braids, links and mapping class group. Annals of Math. Studies 82, Princeton, N.J.: Princeton Univ. Press 1974

BURDE, G., ZIESCHANG, H.: Knots. Berlin-New York: Walter de Gruyter 1985

CROWELL, R.H., FOX, R.H.: Introduction to knot theory. Berlin-Heidelberg-New York: Springer 1977

FENN, R.A.: Techniques of geometric topology. London Math. Soc. Lect. Notes Series 57. Cambridge: Cambridge Univ. Press 1983

HEMPEL, J.: 3-manifolds. Annals of Math. Studies 86, Princeton, N.J.: Princeton Univ. Press 1974

KAUFFMAN, L.: On knots. Princeton, N.J.: Princeton Univ. Press (to appear)

ROLFSEN, D.: [R]

STILLWELL, J.: Classical topology and combinatorial group theory. Berlin-Heidelberg-New York: Springer 1980

Chapter One

HOPF, H.: Über die Abbildungen der dreidimensionalen Sphäre auf die Kugelfläche. Math. Ann. 104, 637-665 (1931)

SEIFERT, H.: [S]

Chapter Two

BONAHON, F., SIEBENMANN, L.: [BS]

DUMBAR, W.: [Du1]

HILBERT, D., COHN-VOSSEN, S.: [HC]

MILNOR, J.: On the 3-dimensional Brieskorn manifolds M(p,q,r).
Annals of Math. Studies 84, Princeton, N.J.: Princeton Univ. Press
1974

MONTESINOS, J.M.: [Mo]

ZIESCHANG, H., VOGT, E., COLDEWEY, H.-D.: Surfaces and planar
discontinuous groups. Lect. Notes in Math. 835. Berlin-Heidelberg-
New York: Springer 1970

Appendix A

BONAHON, F., SIEBENMANN, L.: [BS]

DUMBAR, W.: [Du1]

SCOTT, P.: [Sc]

THURSTON, W.: [Th2]

Chapter Three

DUMBAR, W.: [Du1]

DU VAL, P.: [DuV]

LAMOTKE, K.: Regular solids and isolated singularities. Vieweg
Advanced Lectures in Math. Braunschweig-Wiesbaden: Friedr. Vieweg
u, Sohn 1986

SCOTT, P.: [Sc]

222

SLODOWY, P.: [Sl]

THRELFALL, W., SEIFERT, H.: [TS]

WOLF, J.A.: Spaces of constant curvature. Berkeley, Cal.: Publish or Perish 1977

Chapter Four

BONAHON, F., SIEBENMANN, L.: [BS]

KULKARNI, R., RAYMOND, F.: 3-dimensional Lorentz Space-forms and Seifert fiber spaces (preprint)

MONTESINOS, J.M.: [Mo]

ORLIK, P.: Seifert manifolds. Lect. Notes in Math. 291. Berlin-Heidelberg-New York: Springer 1972

RAYMOND, F.: Classification of the actions of the circle on 3-manifolds. Trans. Amer. Math. Soc. 131, 51-78 (1968)

RAYMOND, F., VASQUEZ, A.T.: [RV]

SEIFERT, H.: [S]

THRELFALL, W., SEIFERT, H.: [TS]

WALDHAUSEN, F.: [Wa]

Chapter Five

ALLING, N.L., GREENLEAF, N.: Foundations of the theory of Klein surfaces. Lect. Notes in Math. 219. Berlin-Heidelberg-New York: Springer 1971

BEARDON, A.F.: A primer on Riemannian Surfaces. London Math. Soc. Lect. Notes Series 78. Cambridge: Cambridge Univ. Press 1984

FORSTER, O.: Lectures on Riemann surfaces. Berlin-Heidelberg-New York: Springer 1981

MAGNUS, W.: [Mag]

Appendix B

BEARDON, A.F.: [Be]

DuVAL, P.: [DuV]

MILNOR, J.: [Mi]

THURSTON, W.: [Th2]

NOTES TO PLATE I

1. <u>*T*</u>. *Frieze of decorative ceramic. Nasrite art, second half of the 14th century.*
    Court of the Arrayanes. (Alhambra).

    Remarks. (i) If this piece is considered as a black and white design, the
    group is S333. The colors lower the symmetry to T.
    (ii) If we look at MA 2227 (*), a fragment of an arch, Nasrite art, 14th
    century in the Lions' Courtyard (Alhambra), we see an example of symmetry T
    and with no question about colors.

2. <u>*K*</u>. *Painting. Nasrite art, 14th century. Gate of the Wine (Alhambra).*

    Remarks. This is a very common group in the Alhambra. However, most of the
    examples found have arabic letters or floral ornaments which reduce the
    symmetry to T (cf. Remark (ii) to D22 in 6. below).

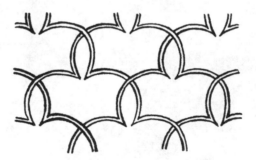

3. *A. MA 1375. Frieze of decorative ceramic. Nasrite art, 14th century*
    (Alhambra).

4. *M. MA 1694. Upper part of a frieze of decorative ceramic. Nasrite art,*
    15th century (Alhambra).

5. *S2222. Plasterwork. Nasrite art, first half of the 14th century. North*
    pavillon of the Generalife (Alhambra).

    Remarks. (i) The ceiling of the room at the right entrance of the façade of
    Comares has exactly the same design as Photo Nr. 2 and belongs to the group
    S2222. It is also Nasrite art of the 14th century.

(*) *MA means "Museo Nacional de Arte Hispano-Musulmán", Palacio de Carlos V,*
     La Alhambra (Granada, Spain). The number after MA refers to the number of
     the piece in the Museum.

(ii) The frieze of decorative ceramic MA 1361 of the Alhambra is Nasrite art of the 14th century. If we ignore the colors (which appear in a disordered way) we classify it as S2222. Here Remark (i) to S333 (in 10. below) applies. Note also that the ground-symmetry group (cf. the Final Comments) is S2222.

(iii) The tile MA 3774 from the store-room of MA belongs to the Convent of St. Francis of the Alhambra. It is Nasrite art of the 14th century and it is classified as S2222.

6. <u>D22.</u> *Painting. Nasrite art, 14th century. Hall of the Kings (Alhambra).*

    <u>Remarks.</u> *(i) The author has found in the Alhambra only two different patterns having symmetry D22. One pattern is a zig-zag and appears in paintings decorating columns - for instance in the Hall of the Kings, in the façade of Comares, in the upper storey of the baths of the Royal House, in the Room of Baraka, and in two ceramic pavings (original or following the original) of the second half of the 14th century in the Hall of the Abencerrajes. The second pattern appears in 12 capitals of the pavilions in the Court of the Lions. There is a painted copy of this design in the upper storey of the baths of the Royal House, and it appears also in the 13th century ceramic vessel MA 1462.*

    (ii) In the pavings of the Hall of Abencerrajes just mentioned, one must assume that if two tiles of the same color touch along a common edge, this edge is disregarded, so that the two tiles form a unit. This certainly was the aim of the artist in designing this paving (*). However it is not always the case: cf. Remark (i) to S632 in 16. below.

7. <u>P22.</u> *Fishbone work. Nasrite art, beginning of the 14th century. Vault of the Gate of the Wine (Alhambra).*

    <u>Remark.</u> *The author has found in the Alhambra only one pattern having symmetry P22. It appears also in the entrance arch of the Tower of Justice, and in many pavings of the Alhambra.*

8. <u>D2$\overline{2}\overline{2}$.</u> *MA 188. Frieze of decorative ceramic. Nasrite art, 14th century. Room of Baraka (Alhambra).*

    <u>Remark.</u> *This is one of the commonest groups in the Alhambra.*

9. <u>D$\overline{2}\overline{2}\overline{2}\overline{2}$.</u> *MA 1362. Paving. Nasrite art, 14th century. Between the Room of Baraka and the Room of the Ambassadors (Alhambra).*

    <u>Remark.</u> *This is also a common group. This particular piece was originally a colored ceramic paving. The colors have almost disappeared. In cases like this, we disregard the colors entirely to classify the pattern, because what remains was certainly well known to the artist. MA 4587 is a different, faultless example of this group.*

(*) *If the tiles are distinguished from one another, this paving is K, but we consider this as an accidental occurrence. In the paper "La simetría y la composición de los tracistas musulmanes" (Investigación y progreso, año VI, n° 3, marzo 1932, pp. 33-45), Antonio Prieto Vives, its author, says: "very often the ornaments and colors follow a part of the symmetry of the lines, in some cases hiding the previous geometric work, as if they wanted to propose an enigma, and this leads one to suspect that the dilettanti of the period derived enjoyment from difficulty in understanding, and this is, no doubt, the reason for the attraction this art holds for those who have penetrated, even in a small part, its secrets".*

10. S333. *Frieze of decorative ceramic. Nasrite art, 14th century. First tepidarium; baths (Alhambra).*

Remarks. (i) To classify this piece we have assumed that it is a black and white design, i.e. we have identified all colors with one another (or with black). The reason is that the colors are distributed in a way that reduces the symmetry to the trivial group, and one has to recognize that the artist was certainly acquainted with both the symmetry concept of the colored pattern and that of the black and white pattern, and suppose that the distribution of colors was made either randomly or for the sake of some kind of harmony defying mathematization. Moreover, if only a reduced and uniformly colored part of a frieze had remained up to our times - as happens with Photo Nr. 16 - we would not have hesitated to classify the frieze as if the original (intact) piece were uniformly colored. This remark, and also the one concerning DĒĒĒĒ, justify adopting liberal rules when dealing with the classification of ornaments. Nevertheless in all the questionable patterns we give alternative instances showing faultless examples of the groups under study.

(ii) An example of S333 in which no problem with colors arises is Photo Nr. 18 which shows the tile MA 1295; Nasrite art, 14th century (Alhambra). The next Figure is a reconstruction of this pattern by the author.

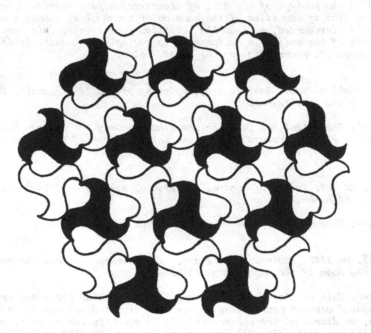

11. D3̄3̄. *Painting. Nasrite art, year 1362. Vault of the Gate of the Wine (Alhambra).*

Remark. The author knows in the Alhambra only one pattern with group D3̄3̄ . Another example of this pattern appears between the Room of Comares and the Court of Arrayanes; it looks like a frieze pattern, but it is big enough to exhibit independent translations. However, if the colors in Photo Nr. 18 are disregarded, it belongs to D3̄3̄, forming a different pattern.

12. $D\overline{333}$. *Plasterwork. Nasrite art, 14th century. Decoration of the "mucarna" in the entrance to the Hall of the Abencerrajes (Alhambra).*

   Remarks. (i) Apparently this is the only example of this group to be found on the building of the Alhambra. However a further example is seen in MA 3113, a chair of "Taracea" work (Nasrite art, 14th century) which is part of the furniture of the 14th century Alhambra (Photo Nr. 19). The group $D\overline{333}$ appears in the chessboard work of equilateral triangles colored black and white, decorating the interior of little circles.
   (ii) The chessboard of equilateral triangles appears also in the exterior wall of Saint Michael's chapel in The "Seo" (Zaragoza, Spain). This is "mudéjar" art of the 14th century following the tradition of Sevilla.

   (iii) The same design appears in a window of "Taracea" work belonging to the MA, originating from a Granada palace. This is Nasrite art of the 15th century. Other examples are two boxes of "Taracea" in the MA and in the Archeological Museum of Granada, but of unknown procedence.

13. $\underline{S442}$. *MA 1359. Fragment of a frieze of decorative ceramic. Nasrite art, 13th century. Generalife (Alhambra).*

14. $D4\overline{2}$. *Fragment of a frieze of decorative ceramic. Nasrite art 14th century. Tower of the Ladies (Alhambra).*

   Remark. Here again we apply the Remark (i) of S333, since the colors are not important. However there are in the Alhambra many instances of this pattern in which no question with colors arises: for instance the design of the doors of Comares, or a plasterwork in the Tower of the Ladies. This one, like a good deal of other plasterwork, might have lost its original colors (cf. Remark in $D\overline{2222}$).

15. $D\overline{442}$. *Fragment of a frieze of decorative ceramic. Nasrite art, 14th century. Tower of the Ladies (Alhambra).*

16. $S632$. *MA 4610. Frieze of decorative ceramic. Nasrite art, 15th century (Alhambra).*

   Remarks. (i) Here all tiles have the same color (the Moors referred to it as "honey-like"). In contrast with Remark (ii) to D22, we do not disregard the lines between tiles in this example because this was clearly the aim of the artist (otherwise, he would have used much simpler tiles).

   (ii) This is a very common group in the Alhambra.

17. $D\overline{632}$. *Plasterwork in window. Nasrite art, 14th or 15th century. Court of the Arrayanes (Alhambra).*

18. $\underline{S333}$ or $D\overline{33}$. *See Remarks to 10. and 11.*

19. $\underline{D333}$. *See Remark to 12.*

Final comments

*The author visited the restauration workshop in the Alhambra. Antonio Molina
Gualda who is in charge of this interesting work said that they follow the same
technique the Moors used: they draw the design on paper, determine the size of
the tiles and colors and finally fire them. It then becomes clear that answering
the question of "what groups the Alhambra builders knew" is not just a matter
of counting the final ornaments on the walls in the Alhambra: one just cannot
ignore the preceding work. Thus in counting ornaments one has to consider the
tiles without colors and with colors; the plasterwork with and without the floral
ornaments, etc. ...*

*The author thinks that the "ground symmetry" (no colors, no floral ornaments, no
lettering, etc. ..) is more representative of the islamic knowledge of symmetry
than the final product. In this line of thought, note that all 17 groups are
certainly represented in the Alhambra, but it seems that they are not all represented as "ground-symmetry" groups. For instance, as far as I know, $D\overline{333}$ is
not represented as ground-symmetry group, indicating a poor development of this
symmetry group.*

Bibliographical remark

*It has been said that only 13 groups were present in the Alhambra (see, for
instance, B. Grünbaum, Z. Grünbaum and G.C. Shepard, "Symmetry in Moorish and
other ornaments", Comp. and Maths. with Appls., 12B (1986) 641-653). Actually,
it is not difficult to find 16. The credit for the detection of the elusive
$D\overline{333}$ (Photo Nr. 12) is to be given to Rafael Pérez Gómez (see Rafael Pérez Gómez:
"The recognition of the plane crystallographic groups, p2, pg, pgg, p3m1, in the
Alhambra", Abstracts of the Amer. Math. Soc. (to appear in 1987)). The example
in Photo Nr. 19 was shown to the author by Prof. Antonio Fernández Puertas.
However, the example of Pérez Gómez is better since it belongs to the fabric of
the Alhambra.*

*Acknowledgements. The photographs were made by Manuel Valdivieso. Professor
Antonio Fernández Puertas pointed out to me the examples of $D\overline{333}$ referred to in
12., Remarks (i) and (iii), while the one in the Abencerrajes and the painting K
in 2., was pointed out to me by Rafael Pérez Gómez, whose knowledge of the
Alhambra from the point of view of ornaments is remarkable. Professor Fernández
Puertas gave me every facility to study and photograph the specimens shown in
the Museum of Hispano-Musulmán art of which he is the Director. He also informed
me of the dating of the different pieces presented in this book. I therefore am
very indebted to him and to Professor Pura Marinetto for sharing with me their
impressive knowledge of Islamic art. Finally I am indebted to the "Patronato"
of the Alhambra for giving me permission to take photographs in the Alhambra.*

Equivalence of notations

Γ	International Notation (short-hand version)	Γ_∞
T	p1	0
K	pg	D_2
A	p1m1 (pm)	D_2
M	cm	D_2
S2222	p211 (p2)	C_2
D22	pmg	D_4
P22	pgg	D_4
D2$\overline{2}\overline{2}$	cmm	D_4
D$\overline{2}\overline{2}\overline{2}\overline{2}$	pmm	D_4
S333	p3	C_3
D3$\overline{3}$	p31m	D_6
D$\overline{3}\overline{3}\overline{3}$	p3m1	D_6
S442	p4	C_4
D4$\overline{2}$	p4gm (p4g)	D_8
D$\overline{4}\overline{4}2$	p4mm (p4m)	D_8
S632	p6	C_6
D$\overline{6}\overline{3}\overline{2}$	p6mm (p6m)	D_{12}

NOTES TO PLATE II

1. *WULFENITE. PbMoO$_4$. The specimen comes from Oujda (Morocco) and belongs to the collection of the "Instituto Geológico y Minero", Madrid (Spain). The photograph, made by Luis Arancón, shows the group S44. The crystals are tetragonal pyramids. The lack of reflecting planes is evident when one looks at the bases of the pyramids (especially the big crystal on the lower right of the photograph).*

2. *AXINITE. (Ca, Fe^{2+}, Mn)$_3$ Al$_2$(BO$_3$)(Si$_4$O$_{12}$)(OH). La Selle, Oisans (France). "Museo Nacional de Ciencias", Madrid (Spain). The photograph, made by Mariano Bautista and Angel Sanz, shows the group P.*

3. *GYPSUM. CaSO$_4$·2H$_2$O. The tabular crystal comes from Portalrubio, Teruel (Spain) and belongs (as do specimens 4 - 10) to the "Instituto Geológico y Minero", Madrid (Spain). The photographs of this and the rest of the specimens were made by Luis Arancón. The group is D2.*

4. *IRON TOURMALINE (SCHORL). (Na,Ca)(Li,Mg,Al)$_3$(Al,Fe,Mn)$_6$(OH)$_4$(BO$_3$)$_3${Si$_6$O$_{18}$}. Guadarrama, Madrid (Spain). The group is D3$\overline{3}$.*

5a,b. *QUARTZ. SiO$_2$. These two interesting, and almost identical, crystals come from Caldas de Oviedo (Spain). The group is S322. Each crystal is formed by two rhombohedra. These two are equally developed in the brown crystal, which looks like a hexagonal dipyramid. The photograph does not show the absence of reflecting planes due to the lack of a trigonal trapezohedron.*

6. *CALCITE. CaCO$_3$. Picos de Europa (Spain). The group is D2$\overline{3}$. The crystal is striated and this shows that the reflecting planes do not contain the binary axes. The crystal is formed by an hexagonal scalenohedron, a hexagonal prism and two rhombohedra.*

7. *SULPHUR. S. Cattalnisetta (Italy). The group is D$\overline{2}\overline{2}\overline{2}$. The crystals show rhombic dipyramids and pinacoids.*

8. *PYRITE. FeS$_2$. Quiravilca (Perú). The group is D3$\overline{2}$ as shown by the striated cubes.*

9. *MAGNETITE. Fe$_3$O$_4$. Brosso Piemonte (Italy). The group is D$\overline{4}$3$\overline{2}$. The octahedron is pyrite. The magnetite crystals are black rhombododecahedra.*

Remark. The photographs illustrate the groups Smm, Pm, Dm, D$\overline{m}\overline{m}$; Sm22, D2$\overline{m}$, D$\overline{m}\overline{2}\overline{2}$; D3$\overline{2}$, D$\overline{4}3\overline{2}$. Missing are S332 (cobaltite and ullmanite, of which good crystals are rare), D$\overline{3}$3$\overline{2}$ (sphalerite), S432 (without undoubted representatives) and S532, D$\overline{5}$3$\overline{2}$ (which cannot occur in minerals).

V. V. Nikulin, I. R. Shafarevich

Geometries and Groups

Translated from the Russian by M. Reid

Universitext

1987. 178 diagrams. Approx. 265 pages. (Springer Series in Soviet Mathematics).
ISBN 3-540-15281-4

Contents: Forming geometrical intuition; statement of the main problem. – The theory of 2-dimensional locally Euclidean geometries. – Generalisations and applications. – Geometries on the torus, complex numbers and Lobachevsky geometry. – Historical remarks. – List of notation. – Index.

This is a quite exceptional book in which an important field of mathematics is treated in a masterly manner that is both lively and approachable. Each stage of the development is carefully motivated by discussion of physical, general scientific and also philosophical implications of the mathematical argument.
The book will hopefully soon be one of the ever-lasting classics like Hilbert and Cohn-Vossen "Geometry and the Imagination" or Hermann Weyl "Symmetry".

Springer-Verlag
Berlin Heidelberg New York
London Paris Tokyo

Springer

M. Berger

Geometry I

Translated from the French by M. Cole and S. Levy
Universitext
1987. 426 figures. XIII, 427 pages. ISBN 3-540-11658-3

M. Berger

Geometry II

Translated from the French by M. Cole and S. Levy
Universitext
1987. 364 figures. X, 407 pages. ISBN 3-540-17015-4

This two-volume textbook is the long-awaited translation of the French book "Géometrie" originally published in five volumes. It gives a detailed treatment of geometry in the classical sense.

An attractive characteristic of the book, and of Prof. Berger's writing in general, is that it appeals systematically to the reader's intuition and vision, and systematically illustrate the mathematical text with many figures (a practice which has fallen into disuse in more recent years).

For each topic the author presents a theorem that is esthetically pleasing and easily stated – even though the proof of the same theorem may be quite hard and concealed. Many open problems and references to modern literature are given.

The third principal characteristic of the book is that it provides a comprehensive and unified reference source for the field of geometry in all its subfields and ramifications, including, in particular the following topics: crystallographic groups: affine, Euclidean and non-Euclidean spherical and hyperbolic geometries, projective geometry, perspective and projective completion of an affine geometry, cross-ratio; geometry of triangles, tetrahedron circles, spheres; convex sets and convex polyhedrons, regular polyhedrons, isoperimetric inequality; conic sections and quadrics from the affine, Euclidean and projective viewpoints.

A companion volume of exercises in geometry has already been published by Springer-Verlag in its "Problem Book" series.

Springer-Verlag
Berlin Heidelberg New Yoork
London Paris Tokyo

Springer

E R R A T A

José M. Montesinos

Classical Tessellations and Three-Manifolds

ISBN-13:978-3-540-15291-0

1. Please replace pp 210, 211, 212 by the attached pp 210',
   211', 212'.

2. Plate I appears on pages 96 and 97 (not on page 95),
   Plate II on page 95 (not on pages 96-97).

ERRATA

José M. Montesinos

Classical Tessellations and Three-Manifolds

ISBN 3-540-15291-0

1. Please replace pp 270, 271, 210 by the attached pp 270, 271, 210.

2. Plate I appears on pages 96 and 97 (not on page 98), Plate II on page 95 (not on pages 96-97).